Interaction of Cells with Natural and Foreign Surfaces

Interaction of Cells with Natural and Foreign Surfaces

Edited by
N. Crawford
and
D. E. M. Taylor

Royal College of Surgeons of England
London, England

Plenum Press • New York and London

Library of Congress Cataloging in Publication Data

Institute of Basic Medical Sciences Symposium on Interaction of Cells with Natural and
Foreign Surfaces (1984: Royal College of Surgeons of England.)
  Interaction of cells with natural and foreign surfaces.

  "Proceedings of an Institute of Basic Medical Sciences Symposium on Interaction of
Cells with Natural and Foreign Surfaces, held November 28–30, 1984, at the Royal
College of Surgeons of England, London, England"—T.p. verso.
  Includes bibliographies and index.
  1. Cell interaction—Congresses. 2. Biological interfaces—Congresses. 3. Cell mem-
branes—Congresses. 4. Biocompatibility—Congresses. 5. Biomedical materials—
Congresses. I. Crawford, Nigel MacRae. II. Taylor, D. E. M. (David Ernest Meguyer)
III. Royal College of Surgeons of England. Institute of Basic Medical Sciences. IV.
Title.
QH604.2.I54   1984   574.87'5   86-16953

ISBN-13:978-1-4612-9307-1   e-ISBN-13:978-1-4613-2229-0
DOI:10.1007/978-1-4613-2229-0

Proceedings of an Institute of Basic Medical Sciences Symposium on Interaction
of Cells with Natural and Foreign Surfaces, held November 28–30, 1984, at the
Royal College of Surgeons of England, London, England

It would be difficult to think of a more important field of
study in medicine than the interaction of cells with natural and
foreign surfaces, but it is rare for those in the practice of
medicine and surgery to get together with the basic scientists to
discuss what they know about it, so that ground of common interest
can be explored. Perhaps the symposium should have been entitled
"Interactions between those interested in cell surfaces..."
because it is here that the chief value of such meetings must lie.
Even a brief perusal of this volume shows just how far advanced
thought and work are on some facets of the topic.

Some of the most pressing problems in medicine are those of
tumour spread and of thrombosis of blood vessels where surface
phenomena are of paramount importance. Our understanding is still
very limited but, as the symposium illustrated, the answers are
only to be found by study at fundamental levels, the work gaining
direction and point from interaction with surgeons and others
concerned with the actual disease states.

One area where interactive research is of great importance is
in the study of prosthetic materials. Although implantation of
foreign materials is an everyday occurrence in hospitals and has
been under clinical study ever since the first suture materials
were used for wounds, the replacement of parts of the body by
prostheses is one of the real growth areas of modern surgery. One
cannot conceive of any significant advance in this being made by
trial and error and the fundamental scientists who presented some
of their work at this symposium demonstrated something of the vast
background of scientific information related to such development.
Without the interaction of such meetings the scientists themselves
would seem to be working relatively in vacuo, for the surgeons are
waiting with their already well developed practical skills to put
to good use any new knowledge. For example, how great would be the
alleviation of suffering if the problems associated with the
implantation of small bore vascular prostheses could be overcome,
for this would open new doors to vascular surgeons for replacement
of small but vital diseased vessels.

Symposia such as this one increase the rate of acquisition of knowledge in a seminal way. Professor Crawford and Taylor are to be congratulated for their far-sighted design and excellent organisation of this meeting. This resulting volume will be a milestone in the progress of medicine. It is most fitting that the symposium was held in the Royal College of Surgeons of England.

                                        Professor Sir Gordon Robson, CBE
                                        Master of the Hunterian Institute
February 1986   Royal College of Surgeons of England

This volume brings together papers based on most of the major
contributions to a Symposium held at the Institute of Basic
Medical Sciences at the Royal College of Surgeons in November 1984.
The topic of this Symposium was conceived by the editors
specifically to try to bring together cell and molecular biologists,
materials scientists, biomedical engineers, pathologists and
surgeons with interests in prosthetic devices and extracorporeal
circuitry so as to focus upon the many morphological and physico-
chemical considerations inherent in cell to cell associations and
cell to foreign and natural surface interactions.

For such an interdisciplinary forum to succeed even halfway
it was imperative that all contributors understood in advance the
heterogeneity of interests of the audience and that they sought
interface areas where some of the problems of cellular interactions,
of biotolerance and biocompatibility could be better defined and
solutions sought through interplay of the diverse knowledge and
experience of the participants present. The same principles were
applied to the preparation of the manuscripts. It is a tribute to
all the contributors that, without exception, they responded to
this challenge with papers which not only adequately set the scene
for their own special interests, but also described their
investigative work which was of a high scientific quality. This
in turn generated much useful discussion and comments which often
continued long after the scientific sessions and intruded into the
social intercourse associated with the meeting.

One might in fact make the observation that it is perhaps
through medical/scientific interfaces that developed throughout
the meeting and the potential collaborative links that emerged
between participants, when common interests and goals were
identified, that the success of this meeting should be judged.

It is felt that at the very least the Symposium offered not
only a window into the character of the many interesting problems
yet to be solved in this broad and complex field, but also
hopefully provided some insight into the ways progress might in

future be made. It is hoped that this book will fulfil a similar
function to a much wider audience.

The editors regret that not all the poster contributions
could be included in this report. All were of high scientific
calibre, beautifully presented and many became focal points for
extended discussion sessions during the meeting.

The editors also want to thank the many organisations who
generously provided financial aid to support this meeting. These
are listed overleaf. We are also most grateful for the lecture
room accommodation and other facilities provided for us by the
Royal College of Surgeons and for their financial support for
this I.B.M.S. Symposium.

N. Crawford
D.E.M. Taylor

ACKNOWLEDGEMENT

The editors are most grateful to the following for their kind generosity in financially sponsoring the Symposium which led to this publication.

Cell Surface Research Fund of St. George's Hospital Medical School, London.

Cyanamid of Great Britain Limited

Ethicon Limited, Edinburgh (U.K.)

Geistlich Sons Limited (U.K.)

W. L. Gore & Associates (U.K.) Limited

Johnson & Johnson (U.S.A.)

Medinvent S.A.

Royal College of Surgeons of England, London.

Smith Kline and French Laboratories, Philadelphia (U.S.A.)

CONTENTS

SECTION 1

CELL SURFACE MEMBRANES: MOLECULAR AND ELECTROKINETIC PROPERTIES

SECTION 2

CELL SURFACE INTERACTIONS AND BIOTOLERANCE: BASIC ASPECTS

SECTION 3

CELLULAR INTERACTIONS IN VASCULAR GRAFTS
AND EXTRACORPOREAL CIRCUITRY

SECTION 4

CELL SURFACE INTERACTIONS AND TUMOUR METASTASES

SECTION 5

CELLULAR INTERACTIONS: GRANULOMAS, HEALING WOUNDS, DRESSINGS AND SUTURES

SECTION 6

OTHER CLINICAL ASPECTS OF BIOCOMPATIBILITY

SUMMARY AND CONCLUSIONS

SECTION 1

CELL SURFACE MEMBRANES: MOLECULAR AND ELECTROKINETIC PROPERTIES

STRUCTURAL ORGANISATION AND FUNCTION OF CELL SURFACE MEMBRANES

N. Crawford

Department of Biochemistry
Royal College of Surgeons of England
Lincoln's Inn Fields, London WC2A 3PN

The surface membrane of a cell can be regarded as a phase boundary whose fundamental function is protective with respect to maintaining a fairly constant intracellular environment despite changes taking place in the external milieu. This membrane, whilst permitting selective communication between the interior of the cell and the outside medium, also controls the cell's osmotic, electrical and chemical properties within a narrow range commensurate with the cell's requirements for full metabolic integrity and functional competence. All of these properties depend heavily upon selective transport processes associated with the surface membrane, some of which regulate the movement of ions and charged molecules across the membrane, whilst others are concerned with maintaining the membrane electrical potential and with the passage of nutrients into the cell and unwanted metabolites out of the cell.

Resident within most mammalian cell membranes are also specific molecular entities or clusters of molecules which act as binding sites or receptors for hormones and other free ligands. Following binding of the ligand to the cell surface, signals are then generated to the cell interior for the triggering and regulation of a wide variety of intracellular metabolic and mechanochemical processes related to the cell's particular functional specialisation. Special functions such as synthesis and packaging in secreting cells, engulfment of particles in phagocytes, electrical impulse transfer in nerve cells, absorptive mechanisms in cells of the intestinal tract, contraction in muscle cells of the heart and other contractile tissues etc. etc. are all in part regulated by transmembrane signal transduction (usually through the generation of second messenger molecules) so that processes within a cell can be quickly responsive to external demands. Most of the membrane transport

3

complexes and receptor molecules are under genetic control and
involve the participation of proteins (some of which have enzyme
activity), glycoproteins and proteolipid complexes distributed,
usually non-randomly, within the structural matrix of the surface
membrane. Associated also with the peripheral membrane of many

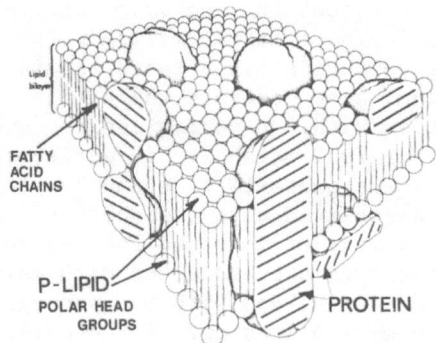

FIGURE 1.   Fluid mosaic model of membrane structure [after
            Singer & Nicholson].

cells are recognition sites involved in cell to cell interactions.
Some of these sites may be linked non-covalently to, or have loosely
absorbed, extracellular proteins such as fibronectin and other
lectins which are capable of generating adhesive or cohesive forces
between the cell and adjacent natural or foreign surfaces.

        Original concepts about the structural organisation of cell
membranes presented a rather static picture, in keeping with the
classical "tramline" images seen in electron microscopy. We now
know, however, that all cell membranes are in a highly dynamic
state and the "fluid mosaic" model presented by Singer & Nichol -
son [1] is now widely accepted by membranologists to be a more
appropriate representation. This model fits better the biophysical
evidence that some of the lipid and protein constituents may migrate
by diffusion and/or rotation within the plane of the membrane.

The Singer-Nicholson concept is illustrated in Figure 1 and it
can be seen that whilst the organisation of the phospholipids as a
bimolecular leaflet has survived from the earlier "unit membrane"
theory of Robertson[2,3], the protein and glycoprotein constituents
are conceived to be embedded within the phospholipid bilayer,
some partially and some transmembrane, spanning both the inner and
other leaflets. The glycoproteins with their covalently attached
carbohydrate moieties are seen as externally oriented complexes
with thier sugar chains extending well beyond the lipid bilayer
region. The significance of this outwardly facing coat known
as the "glycocalyx" is now becoming well recognised in cell surface interactions and further reference to this will be made later
in this paper.

The composition of cell surface membranes does of course vary
somewhat with their origin, but all surface membranes generally
contain around 40% of their dry weight as lipid, approximately 50%
as protein and the carbohydrate present accounting for between 2 and
10% of the dry weight [4]. A small proportion of this carbohydrate is
associated with lipid as glycolipid but the majority is complexed
with protein as glycoprotein. Water represents some 20% of the
total weight of surface membranes and this is tightly bound by
ionic bonding and is vital for the structural and functional integrity of the membrane. Membrane lipids, the majority of which are
phospholipids, are amphipathic; that is their molecular structure
includes a hydrophobic region commonly comprising fatty acids
esterified to the glycerol core and a polar (hydrophilic) head
group consisting of choline, ethanolamine, serine or inositol, esterified through a phosphate linkage to the glycerol backbone. Most
mammalian surface membranes also contain a small proportion of phos-
phosphingolipids in which the molecular region of high hydrophilicity
is due to a phosphate diester comprising sphingosine and phospho-
choline. The commonest of these latter sphingolipids in cell surface
membranes is sphingomyelin and its presence is a distinguishing
feature between surface and intracellular membranes. The hydrophobic
tails of the polar lipids in membranes consist largely of long chain
fatty acids with 16, 18 or 20 carbon atoms, some of which may have
unsaturated bonds. The saturated fatty acids which have C-C bond
angles of 111° pack together well into an ordered integrated zig-zag
formation consistent with the membranes' liquid crystalline structure
but the presence of an unsaturated bond (bond angle 123°) kinks the
chain destroying this order and produces zones of higher fluidity
in the lipid bilayer.

Although the unsaturated fatty acids may range from 16 carbon
atoms with one double bond to 20 carbon atoms with 4 double bonds,
one of the latter, arachidonic acid, is particularly important in
membrane phospholipids as a precursor for a class of pharmacologically active agents known as prostaglandins.

A third class of lipid which is a major feature of cell surface membranes is cholesterol.

This lipid has a compact, mainly hydrophobic structure which has a polar hydroxyl group at one end of the molecule. Surface membranes show a higher content of cholesterol than other membranes and structurally it intercalates between the fatty acid chains of the phospholipids and particularly in regions of unsaturated fatty acids where more space is available. Cholesterol is believed to have a stabilizing role on the membrane and to be strongly influential in determining the fluidity and deformability of blood cell membranes which have to penetrate the microcirculation where the bore of the vessels may be less than their diameter [5]. In circulating blood cells cholesterol is partly exchangeable with lipoprotein cholesterol in the plasmatic environment, but its content within the membrane is maintained at a reasonable constant ratio of approximately 1:1 on a molar basis to the phospholipid by the action of the plasma enzyme lecithin cholesterol-acyl-transferase (LCAT).

Turning to the protein constituents of cell surface membranes, as techniques have improved for preparing surface membranes from cell homogenates in a reasonably purified form and substantially free of contamination by boundary membranes of subcellular organelles, our knowledge of these surface proteins and the part they play in the functional activities associated with the membrane has increased enormously. Many cell surface membrane proteins are glycoproteins in which the carbohydrate moiety is linked to the amino acid back bone of the protein at serine, threonine, asparagine or hydroxylysine residues. The carbohydrate chains may be branched and contain 2-6 sugar residues terminating in fucose, galactose or sialic acid (N-acetyl neuraminic acid). It is this latter sugar derivative that is responsible for a substantial proportion of a cell's surface membrane electronegativity (see Crawford this volume) since it is the only sugar residue present which has a free carboxyl group. The molecular heterogeneity of the sugar units, the degree of chain branching and the different protein linkages possibly all contribute to the specificity of certain membrane glycoproteins with respect to antigenic determinants. The exposed proteins and glycoproteins on the surface of cells are believed to be involved in adhesive phenomena since many different molecular structures are possible through variations in sugar moieties and chain branching. Fibronectin, lectins and other carbohydrate reactive proteins associated with the glycocalyx promote adhesion and spreading of cells on surfaces. The effects are complex, however, since carbohydrates which compete for the lectin binding sites on cells lose their effect if added some minutes after the adhesion-promoting ligand [6] It is believed, therefore, that the primary action of the surface binding agent is to encourage the cells to reorganise their membrane constituents to reveal adhesion sites. Intrinsic membrane glycoproteins are rarely distributed randomly and the possibility that

domain segregaton occurs where adhesion-promoting clusters of glycoproteins are present, held together by hydrophobic interactions with
each other and membrane phospholipids, is receiving considerable
research attention. Such selective segregation, where it has been
studied, has been shown to require full metabolic competence within
the cell and the role of the nucleotides ATP and GTP at the cytoplasmic face must also be taken into account. For a review of these
surface carbohydrates in cell adhesion see Ravvala (1983).

There is now widespread recognition that intracellular cytoskeletal elements play a major role in regulating many properties
of cell surface membranes [7]. Every cell seems to have the capacity
to reorganise its cytoskeletal network by assembly, disassembly,
cross-linking and binding mechanisms. Drugs which block the action
of intracellular cytoskeletal structures and metabolic inhibitors
can diminish the attachment of cells to each other or to foreign
surfaces. The formation of surface adhesion sites is believed
therefore to be also dependent upon the functional integrity of the
microfilament and microtubular elements seen in cells subadjacent
to their surface membrane [8]. Almost all cells display a filamentous
contractile network associated with the surface membrane on the
cytoplasmic face. The major protein constituent of this microfilament network is actin, attached to membrane proteins as fibrillar
actin, bound either directly or through some intermediate connecting
protein. The fine structured submembraneous meshwork of actin
filaments seen in many cells, but particularly well expressed in
cells which undergo motile activities, are a stabilizing influence
on the membrane components and probably attach constraints upon
their freedom of movement within the lipid bilayer. In cells
attached to or moving over a flat surface, areas of the plasma membrane close to the substratum may show a higher level of organisation
of the actin known as "stress fibres". These stress fibres are
organised bundles of filamentous actin side to side associated and
the point of attachment to the membrane on the cytoplasmic face is
known as a "focal adhesion site". The polymerisation and depoly-
merisation of these actin filaments (F-actin) from a subunit pool
reservoir of G-actin are the processes which are believed to effect
cell surface interactions and provide the transitory changes in
membrane cytoskeletal organisation seen in the contact areas during
cell locomotion over surfaces. Immunocytochemistry with antibodies
to known cytoskeleton proteins and interference reflection microscopy
both show a very different level of organisation of the membrane at
these focal adhesion sites than exists in the areas of attachment
of the more usual cortical network of filaments.

Microtubules are a highly ordered arrangement of tubules which
are very prominent in some cells, in the central core of "axonemes"
of cilia and in the mitotic spindle of dividing cells for example,
but in other cells they course through the cytoplasmic matrix without any fixed orientation. There are unfortunately few proven and

well established examples of microtubule interactions with plasma
membranes but studies with the microtubule poisons colchicine and
vinblastine have provided circumstantial evidence that cell surface
receptor topography and the membrane segregatory events seen during
phagocytosis are membrane processes which are in some way affected
by microtubule integrity. Some intrinsic membrane constituents can,
for example, migrate laterally and totally out of a forming phagocytic vesicle. Clearly the structural plasticity of the membrane
juxtaposed to a substratum has changed to a very significant degree.

A large variety of enzyme activities have been identified in
mammalian surface membranes, some of which may be found elsewhere
in the cell whilst others are sufficiently exclusive to the surface
membrane that they can be exploited as markers in membrane isolation
and purification procedures. Additionally, some membrane enzymes
(known as ecto-enzymes) operate on external substrates whilst others
require an internal supply of substrate. An example of the former
class, 5'nucleotidase, is a feature of the surface membranes of
most circulating blood white cells and in the latter class the ion
transport ATPases and adenylate cyclase (which produces the second
messenger cAMP) are good examples. The enzymatic activities of
membrane located enzymes is often greatly influenced by the surrounding environment of other membrane constituents. Some enzymes,
if isolated from the membrane matrix (even by mild procedures) lose
their activity and a particular feature of many of the ATPases is
that activity can frequently be regenerated by the addition of
phospholipids to the isolated enzyme. The role of the lipid in
such instances is generally regarded as due to the reassociation
of different monomeric species to form the active multimeric enzyme
complex. In some circumstances the changes in the character of the
phospholipid fatty acid chains are sufficient to alter the physical
state of the lipid domain closely associated with the enzyme proteins
so as to change the enzymes' kinetic characteristics. Such findings
have encouraged interests in possible dietary modifications to the
surface membranes of cells by modifying their phospholipid fatty
acid composition. For example, the dietary alterations of blood
platelet surface membrane composition and properties to inhibit [9]
platelet aggregation is one area of intensive research activity.
Of the enzymes present in the surface membrane perhaps the most
important group in the context of maintaining the overall integrity
of the cell are the ion transport ATPases whch regulate both the
osmotic and electrical properties of the cells. The major transport
ATPases are the Na^+K^+ ATPases and the $(Ca^{2+} + Mg^{2+})$ ATPase. The former
enzyme, earlier referred to as the "sodium pump" is the major system
for maintaining high intracellular K^+ concentrations and the $Ca^{2+} +$
Mg^{2+} ATPase is involved in the mobilisation of Ca^{2+} either as an
extrusion pump present in the surface membrane of many cells or as
a mechanism for sequestering Ca^{2+} within intracellular membrane
stores [10]. Since in some cells sodium is co-transported into the
cell across the surface membrane with other molecules (amines and

amino acids) the Na^+K^+ ATPase serves to maintain low Na^+ concentrations in the cytosol compartment. Both types of ion transport pumps are energy dependent processes and are capable of operating against unfavourably high gradients of the transported ions. For example, the Ca^{2+} ATPase of erythrocyte membranes operates as an extrusion pump and helps to maintain a cytosol Ca^{2+} level in the submicromolar range even though the concentration of Ca^{2+} in the external plasmatic environment may be 10,000 times greater than the cytosol level. These active transport processes require ATP as substrate for the ATPase and as a phosphate donor for certain protein phosphorylation events. They make high demands upon the metabolic capacity of cells. Other non-energy requiring systems for moving ions across cell membranes are the cation ion exchange processes best illustrated by surface membrane Na^+/Ca^+ counter ion exchange and the voltage gated channels present in the membranes of many cells. Detailed discussion of these ion transfer processes is beyond the scope of this present paper.

There are few fields of biology that have attracted as much research attention as that directed towards understanding the structure and function of cell membranes. Although much progress has been made, the problems that exist in probing the dynamic and thermodynamic aspects of cell membrane aspects are still most formidable. Electron microscopy has allowed us to see membranes as the key interface of a cell with its environment and chromatography, X-ray diffraction, amino acid and carbohydrate sequencing, spin labelling and a host of other technical approaches are beginning to reveal much about membrane constituents, their three-dimensional organisation and 'small print' molecular detail. However, there is still much to do and we are only just beginning to investigate the more dynamic aspects of these structures and the studies of cellular interactions with natural and foreign surfaces as well as cells with each other have been to date largely observational. With all the diversity in composition and function of cell surface membranes it's perhaps surprising that any sort of unifying hypothesis has emerged, but the fluid mosaic model, now widely accepted by membrane enthusiasts, has swept away all other concepts that were restrictive in our understanding of the dynamic interplay that exists between the various molecular entities in the cell interface. Progress in the next few years should be even more rapid than in the past decade.

REFERENCES

1. S.J. Singer and G.R. Nicholson, _Science_, 75, 720–31, (1972).
2. J.D. Robertson, _Biochem.Soc. Symp._ 16, 3–43, (1959).
3. J.D. Robertson, _Symp. Int. Soc. Cell Biol._ 5, 1–59, (1966).
4. R. Harrison and G.G. Lunt, Biological membrane: Their structure and function. Chapter 5 pp 62–107, Blackie, Glasgow and London. (1980).

5.   G. Nicholson, Biochim. Biophys. Acta. 457, 57-108, (1976).

6.   H. Ranvala, Trends Biochem. Sci. 8, 323-325. (1983).

7.   C.M. Cohen and D.K. Smith, in: Enzyme of Biological Membrane, ed. A.N. Martonosi, Vol. 1. pp 29-79, Plenum Press, New York, (1985).

8.   J. Oliver and R.D. Berlin, in:Actin: Structure and Function in Muscle & Non-muscle Cells, ed. C.G. dos Remedios and J.A. Barden, pp 259-266, Academic Press New York, (1982).

9.   S. Renaud, Adv. Exp. Med. Biol. 63, 265-276, (1975).

10. A.N. Martinosi, in: Muscle and Non-muscle Motility Vol.1, Chapter 5, ed: A. Stracher pp 233-357. Academic Press Inc. (1983).

ELECTROKINETIC ASPECTS OF CELL SURFACES

N. Crawford

Department of Biochemistry
Institute of Basic Medical Sciences
Royal College of Surgeons of England
Lincoln's Inn Fields, London, WC2A 3PN

All living cells are surrounded by membranes, which are their major interface with the external environment, and it's only through the special properties of these selective permeability barriers that cells can regulate their intracellular environment within the limits necessary for optimal function. The surface [or plasma] membrane of a living cell also plays very significant roles in a range of fundamental biological processes such as growth, cell division, secretion, motile activities, cell to cell recognition and mitogenicity to name just a few. Moreover, many of the cell's biochemical pathways necessary to maintain metabolic health are controlled by supplies of fuel entering the cell through its surface membrane and by molecular stimuli acting on surface receptors from which a signal is transduced into the cell interior to trigger metabolic and functional processes.

It's not surprising therefore that the characterisation of constituents present in cell membranes, their topographical distribution and dynamic aspects of their organisation has accounted for a very substantial part of the research interests in cell biology during the last 2-3 decades. Although initially attention was focused on the lipids present in a membrane and their bilayer organisation and the role of phospholipids and cholesterol, later studies have included the intrinsic proteins and glycoproteins whose influence at the molecular level may extend some 20-30 nm outside the bilayer. We now also know that many of the physical properties of a membrane, its fluidity and deformability for example, as well as its properties and receptor status depend upon interactions between intrinsic membrane constituents and the cytoskeletal structures and miniature muscle complexes lying subadjacent to a

11

membrane on the cytoplasmic face. The influence of the membranes
here extend deeply into the interior of the cell significantly
affecting important properties and functions.

Some of the more descriptive aspects of cell membranes and their
functions as selective permeability barriers, as well as the document-
ation of major constituents [lipids, proteins and enzymes etc.] have
been dealt with in the preceeding paper at this meeting [1]. In the
present paper emphasis has been placed on the electrokinetic aspects
of cell membranes and how the presence of electrically charged constituents within or associated with the lipid bilayer may influence
a cell's interactions with other cells and with different foreign
and natural surfaces.

POTENTIALS

When water is present, as in all biological systems, an interface between two phases is generally the site of electrical charge
and a potential difference exists across the boundary. Almost all
cell membranes, being semi-permeable barriers, transport ions by
energy dependent pumps or mobilise ions by voltage gated channels.
In this way differential ion gradients are maintained superimposed
above the tendency for the ions to passively diffuse to equilibrium.
Because of this unequal distribution of ions a potential difference
results across the cell boundary membrane which is usually termed
"membrane potential" or "transmembrane potential" (E_m). Under normal
physiological conditions and at equilibrium there are thermodynamic
constraints on the magnitude of this interfacial or membrane potential and in many larger sized cells this potential can be measured
using a penetrating fine-tipped micropipette electrode, with a high
impedance, inserted into the cell and a suitable extracellular electrode in the surrounding medium. Most mammalian cells exhibit transmembrane resting potentials of the order -10 to -30 mv with some
cells reaching values as high as -60 mv [2].

However, although much attention has been focused upon these
transmembrane potentials and particularly they have been extensively
studied in smooth muscle and in cancer cells and other cells during
mitotic division, the differences demonstrated appear to be more due
to variations in the Na^+-pump, causing alteration to transmembrane ion
gradients or in the amount of Ca^{2+} bound to the cytoplasmic face of
the membrane, rather than to the differences in charged molecular
entities intrinsic to the membrane structure. For these reasons
the relevance of different expressions of transmembrane potential
to cell-surface interactions is as yet unclear and microelectrode
measurements with cells in contact with natural or foreign surfaces
have produced equivocal results. However, the electrical property
which does seem to play a major role in such surface interactions is
the net electrical charge expressed at the outer surface of the cell
membrane due to polar groups of intrinsic membrane components

sufficiently surface orientated to allow interactions with the
adjacent ionic environment.

ZETA POTENTIAL AND ELECTROPHORETIC MOBILITY

A living cell suspended in an ionic medium attracts an ion cloud
or electrical double layer [Stern layer] which may extend as much as
10-15 Å outside the cell periphery (Figure 1). The electrical
potential at any point in this ion cloud is a net result of the

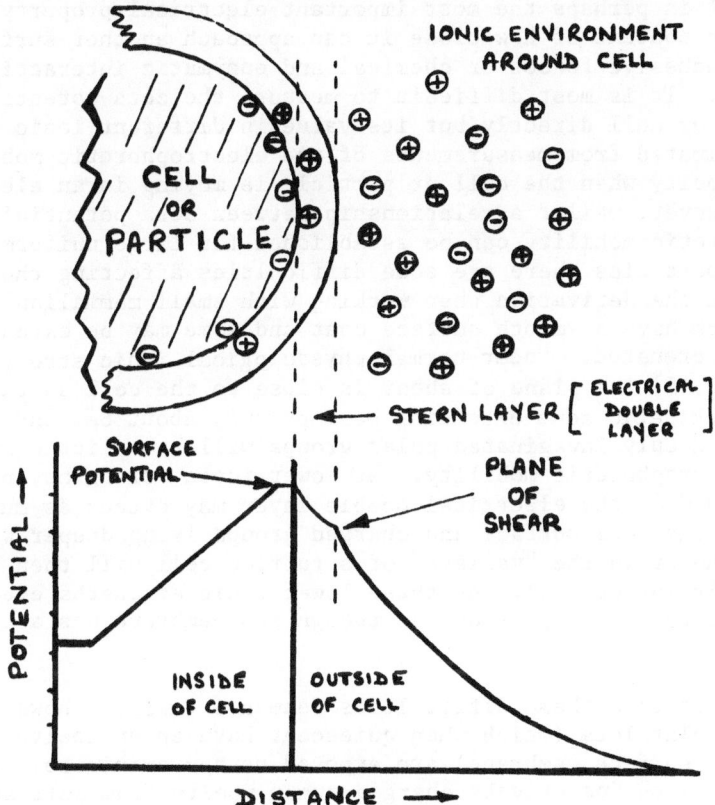

Fig.1. A particle with an overall negative charge attracts cations
        from the medium towards its surface. In consequence some
        anionic groups are neutralised. Thermal fluctuations pre-
        vent total neutralisation and there is a statistical dist-
        ribution of an excess which is greater nearer the particle
        surface. The electrical double layer [Stern layer] is the
        region of charge imbalance and ions touching or very close
        to the particle surface become immobilised. The electrical
        property which determines the particles' electrophoretic
        mobility is the potential at the plane of shear between
        the electrical double layer and the ionic environment.
        This potential is knowm as the "Zeta potential".

charges resident on the surface of the cell and any ions of opposite
charge in the medium which have been attracted towards the cell
surface. This results in a potential which Helmholtz [3] named the
"zeta potential". This electrokinetic property can be regarded
essentially as the potential difference between the plane of shear of
the ionic cloud [or electrokinetic zone] with the ionic environment
and some remote point in the cell suspension medium. This zeta
potential is of course always significantly less than the summation
of the actual charged species on the cell surface due to ion neutralisation but is perhaps the most important electrical property of a
cell in the context of how close it can approach another surface
such that adhesive forces or chemical and enzymatic interactions can
then occur. It is most difficult to measure the zeta potential of
a particle or cell directly but its value in different ionic media
can be estimated from measurements of the electrophoretic mobility,
or the velocity when the cell or particle is moving in an electrical
field. However, whilst a relationship between zeta potential and
electrophoretic mobility can be established for large uniformly
spherical particles there are some difficulties affecting the
accuracy of the derivation when working with small mammalian cells;
few of which have a smooth surface coat and some may be extensively
ruffled or crenated. Under normal physiological ionic strengths
[i.e. $\mu \sim 0.15$] the plane of shear is close to the cell surface with
the electrokinetic zone extending perhaps only about 8Å. Under such
conditions deeply invaginated polar groups will have little influence
on the electrophoretic mobility. At lower ionic strengths, however,
[i.e. $\mu \sim 0.05$], the electrical double layer may extend as much as
15Å beyond the cell surface and charged groups lying deeper within
the membrane or in the "valleys" of a ruffled cell will then be
electrically influential. At these lower ionic strengths even a
highly ruffled cell may behave as though its membrane was a smooth
surface.

To illustrate these difficulties Seaman & Vassar[4] showed that
when blood platelets (which when quiescent have an extensively
invaginated surface membrane) are exposed to neuraminidase, an enzyme
which cleaves the negatively charged sugar moeity from cell surfaces,
much more sialic acid could be removed than the theoretical amount
calculated from the measured reduction in electrophoretic mobility.
Clearly, therefore, not all the negatively charged carboxyl groups
of the sialic acid present are electrokinetically influential at
the plane of shear between the electrical double layer and the
surrounding ionic medium.

A further consideration, which is particularly relevant in
dealing with blood cells suspended in their normal plasmatic environment, or a similar colloidal medium, is the presence of loosely and
non-specifically absorbed proteins. Such a "proteinaceous halo", if
partially or wholly maintained during transfer to the ionic medium
for mobility measurements will significantly influence the cell's

electrokinetic behaviour. The inconsistency in the literature values
for electrophoretic mobility measurements with certain cells is
almost certainly due to differences in these surface absorption
effects.

Despite these reservations, by using low ionic strength conditions, controlled pH and standardised cell preparations, reproducible values for electrophoretic mobilities [usually expressed as
μmetres/second/volt/cm] have been recorded for a wide variety of
mammalian cells. These data are now usually obtained with the use
of analytical electrophoresis units using a laser doppler optical
system which can give mean values with statistical parameters for
as many as 200 or more cells during a single electrophoretic measurement.

The zeta potentials derived from such electrophoretic mobility
measurements for most mammalian cells generally fall within the
range -10 to -30 millivolts.

What then are the surface membrane ionisable groups which contribute to these zeta potentials in living cells? Table 1 shows the
major cationic and anionic groups present in mammalian cell membranes. Despite often substantial charge neutralisation by ions in
the surrounding environment most mammalian cells are electronegative
and one of the major contributors to this is the aforementioned
sialic acid or n-acetyl neuraminic acid (Fig.2). This charged sugar
group is generally the terminal group on some of the carbohydrate
chains of membrane glycoproteins and the charge characteristics
reside in the carboxyl group [-COO'] of the sialic acid moeity.
Although these charged sugars are believed to be the major contributors of negative electrical charge, the α, β and γ COO' groups of
certain amino acids in membrane proteins may also influence electrokinetic behaviour, though generally to a lesser degree. Other
electronegative groups with a potential effect upon cell surface
charge characteristics are phosphate groups and certain other unidentified anionogenic groups. The former have been investigated
with certain cell types by the use of alkaline phosphatase, making

Table 1.   Groups contributing to cell surface charge

Cationic   Anionic

Polypeptide Terminal $-NH_2$ Sialic Acid -COOH
Lys, Hydroxylys $-NH_2$ Carboxyl groups of GLU and ASP
Arginine Guanidine Group   Phosphate groups [proteins and P-
Histidine Imidazole   lipids]
Aminoglycolipids   Sulphate groups
Quaternary Choline
Sulphydryl -SH

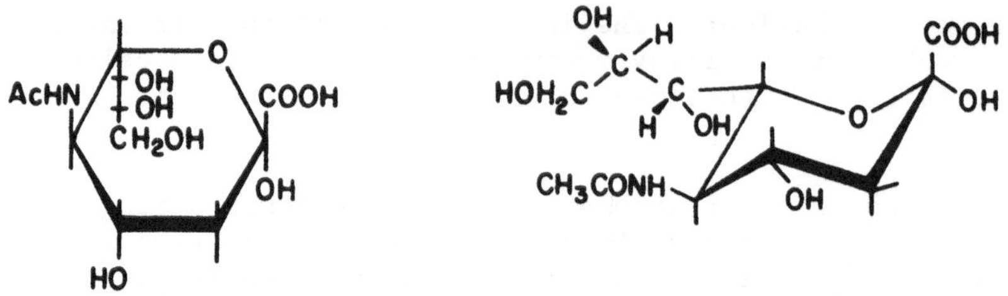

Fig.2   Structural formulae of sialic acid (N-acetyl-neuraminic acid)

electrophoretic mobility measurements before and after treatment with the enzyme, [5,6] but the latter unidentified anionogenic groups have only been roughly determined subtractively after treatment with neuraminidase and phosphatase. The only major cationogenic contributors which are associated with cell surface membranes are the amino groups of intrinsic membrane proteins (containing the amino acids lysine and arginine). These have been shown to increase in experimental studies with trypsin and other proteolytic enzymes and similar protease effects may of course occur in cells in vivo. Titration of such groups with aldehydes and other agents does not, however, markedly change a cell's electrophoretic mobility towards an anode and it is assumed therefore that they have pK values too high to be eletrokinetically influential in the normal physiological range. Other weakly contributory chemical groups present on cell surfaces are the sulphydryl -SH groups. These can be titrated with agents such as dithionicotinic acid but only oxidative processes capable of forming -S-S- bonds within the membrane are likely to influence the electrokinetic behaviour due to conformational changes revealing anionic or cationic groups previously non-contributory.

Many cells in the body are surrounded by free Ca^{2+} ions and there is substantial evidence that the divalent cation can be associated with sialic acid and with the phosphate groups of proteins and phospholipids. Our knowledge of how bound calcium affects a cell's electrokinetic properties is at present rather sparse but the Ca^{2+}-induced configurational changes that may occur in membrane constituents which would affect short range electrostatic repulsive forces could be of major significance in cell/cell or cell/foreign surface interactions. Additionally, the surface binding of 'adhesive proteins' such as fibronectin and other proteins such as fibrinogen and thrombospondin may be promoted by Ca^{2+}-induced surface alterations revealing binding domains. Such domain revelation or segregation and the possible effects on cell/surface interactions has been little investigated to date but fully warrants more intensive studies.

More information is required for example about how the intracellular
cytoskeletal elements control surface membrane topography and what
are the factors which affect the stability of the membrane cytoskeletal network. Many are Ca^{2+} dependent processes.

FACTORS AFFECTING CELL SURFACE INTERACTIONS

     In general, therefore, the surface charge characteristics of a
cell membrane and the capacity of a cell to interact with other cells
or surfaces depends upon a number of different factors, viz. :-
a) The relative properties of positive and negative groups capable
of expressing their charge characteristics at the plane of shear of
the electrokinetic zone.
b) The pKa of the various ionogenic groups
c) The ionic conditions of the surrounding medium
d) The existence of surface membrane domains with "hot spots" of
interactive polar groups
e) non-specifically adsorbed proteins contributing to or masking
charged moeities
f) The fluidity and deformability characteristics of the cell
membrane.

     A feature too of cell/cell and cell/surface interactions about
which we have little understanding are the changes that result
through senescence. There is little doubt that many of the circulating blood cells show wide ranges of sub-population heterogeneity
These heterogeneity profiles may show significant differences in
surface charge characteristics as well as in metabolic health and
functional competence. The extent to which these various properties
are senescence-related is at present not known but clearly these
differences will also be reflected in cellular interactions with
foreign and natural surfaces. Studies with blood granulocytes and
their more electronegative precursor cells in the bone marrow have
shown that motile properties involving surface events such as
spreading, locomotion and phagocytosis are all better expressed in
the less electronegative subpopulations.

     Such relationships also suggest that membrane constituents
concerned with surface electrical charge expressions [e.g. sial-
glycoproteins and glycolipids] may exercise some form of transmembrane control over a cell's cytoskeletal or miniature muscle equipment since these structures are known from drug studies to be oper-
ationally important in membrane mediated cell motile phenomena.

     The possible importance of all these as well as other factors
such as mutual adhesiveness, contact inhibition etc. in the way
cellular interactions occur should stimulate intensive studies in
these areas which are the basis of so many life processes. It is
feared, however, that little, if any, significant advances will be
made in any of these areas until new techniques emerge for invest-

igating the more dynamic aspects of the microarchitecture of cell
surface constituents at the molecular level. Electrically influential constituents can then be identified and their contributions
measured as interactive dynamic changes occur.

ADHESION-DEPENDENT CELL SURFACE INTERACTIONS

        Intercellular adhesion is a fundamental biological property
which is known to play a key role in processes such as embryonic
differentiation, tissue regeneration, haemostasis and thrombosis,
in metastatic tumour spread, as well as in many other important
cellular phenomena. Over the years there have been many theories
offered to explain adhesion dependent cell/cell and cell/surface
contact processes. As our knowledge of the cell surface membrane
has increased at the molecular level none of these theories have
survived as a unifying hypothesis and the highly dynamic nature of
most cell membranes has meant that many different factors have now
to be taken into account. Concepts appropriate to different surface interactions are unlikely, however, to be mutually exclusive.

        One of the more attractive theories [7] proposed the role of
divalent cations [principally Ca^{2+} and Mg^{2+}] in binding to negatively
charged groups in the membranes and forming bridges with an adjacent
surface. Although the use of EDTA and other chelating agents is an
established procedure for dissociating cells there is little evidence for bridge formation being the major mechano-chemical linkage
between adherent cells. The presence of the divalent cation in
lowering the negative charge and thus decreasing electrostatic
repulsion has also been substantially discounted as far too simplistic. In fact cell aggregation studies, where different divalent cations are used at concentrations adjusted to bring the negative charge
down to the same level, have shown that magnesium is a much more
effective cation for promoting aggregation than calcium, and the cations barium and strontium have little, if any, aggregatory effect [8].
A role for external Ca^{2+} has however been demonstrated on blood platelets in the binding of fibrinogen to the platelet surface. Fibrinogen is an adjunct to platelet/platelet interactions and binds to a
glycoprotein complex consisting of two closely associated molecular
entities GPIIb and GPIIIa. The complex formation appears to be a
calcium dependent event and the two membrane proteins dissociate in
the presence of chelating agents [9]. Although it is possible that
Ca^{2+} ions can participate intimately in some sort of adhesive bond
formation between cells this is more likely to be by their action
in inducing topographical changes of membrane constituents than by
simply reducing electrostatic repulsion between the surfaces.
Another concept earlier proposed was that molecular bridges can be
formed through interactions with molecules in the surface coats and
extruded or released components from within the cells [10,11]. This
theory was strengthened by the reports that the synthesis of certain
components found in exudates correlated with increased adhesiv-

ity [12,13] . Molecular bridging however requires a minimum surface to surface distance [thought to be ∼10Å] when the energy level is then compatable with the formation of chemical or electrostatic bonds or enzyme-substrate complexes. Beyond this distance the attractive forces between cells are substantially less than the forces of repulsion due to the radius of curvature of two approaching cells. This repulsive barrier of course alters when one is dealing with cells and flat surfaces such as is the case with blood cells passing through extracorporeal circuitry or cell associations with implant materials 'in vivo' . The current interest in adhesion promoting molecules such as fibronectin and Von Willibrand factor protein is to some extent due to a revival of the molecular-bridging theory. The finding that fibronectin has a specific collagen-binding domain near the amino terminal led to the concept that secreted fibronectin was the platelet collagen receptor involved in heamostatic processes [14,15] . This cell/surface interaction is however clearly more complicated than can be accounted for by platelet/fibronectin/collagen molecular bridging. Another platelet secreted protein which remains closely associated with the suface membrane is thrombospondin. Thrombospondin not only has binding sites for fibrinogen but direct cross-linking between collagen and thrombospondin has been shown to occur [16]. Separating pure electrokinetic aspects of cell to cell or cell to foreign surface interactions from the variety of bridge-forming chemical processes that can occur as well as covalent cross link formation will be a very formidable task indeed. Further comments on surface glycoproteins as binding sites for ligands which affect adhesivity will also be found in an accompanying paper in these symposium proceedings [1] .

In many clinical studies we are now having to concern ourselves more with the bioelectrical nature of living cells and the surfaces they may interact with. We may on the one hand wish to inhibit close encounters of cells with sufaces as in extracorporeal circuitry, blood and graft storage etc. whilst on the other hand with blood vessel prostheses and wound healing materials, a more satisfactory marriage between "man-made" and "god-made" surfaces is the general goal. The increasingly closer cooperation of clinicians, bioengineers, cell biologists and material scientists etc. augers well for future advances in this field, but only if all take full cognisance of each other's disciplines and their limitations so that objectives are always in focus with the practicalities.

REFERENCES

1. N. Crawford, Accompanying paper in these proceedings, (1986).
2. D.E. Maslow and L. Weiss, Exp. Cell Res., 71, 204-209 (1972).
3. H. Helmholz, Weid. Ann. 7 337-347 (1879).
4. G.V.F. Seaman and P.S. Vassar, Changes in the electrokinetic properties of platelets during aggregation, Arch. Biochem. Biophys., 117, 10-17.

5.   J.N. Merishi, Phosphate groups (receptors) on the surface of
      human blood platelets, Nature, 226, 452–453 (1970).
6.   J.N. Merishi, Molecular aspects of the mammalian cell surface
      in Prog. Biophys. Mol. Biol., 25, 1–70, ed. J.A.V. Butler
      and D. Noble Pergammon Press, Oxford (1972).
7.   B.A. Pethica, Exp. Cell Res. Supp., 8, 123–142 (1961).
8.   P.B. Armstrong, J. Exp.Zool., 163, 99–107 (1966).
9.   D.R. Phillips in "Platelet Membrane Glycoproteins" (J.N. George
      A.T. Nurden and D.R. Phillips eds.) pp145–169, Plenium Press
      New York (1985).
10.   A. Rambourg, Int. Rev. Cytol., 31, 57–68 (1972).
11.   R.J. Winzler, Int. Rev. Cytol., 29, 77–84 (1970).
12.   D.E. Maslow and L. Weiss, Exp.Cell Res., 71, 204–209 (1972)
13.   G. Poste, Exp. Cell Res., 77, 264–276 (1970).
14.   E. Ruoslahti, E.G. Hayman, P. Kuusela, J.E. Shively and E.
      Engvall, J. Biol. Chem., 254, 6054–6059 (1979).
15.   M. Firie and D.B. Rifkin, J. Biol. Chem., 255, 3134–3140 (1980).
16.   J. Lahav, M.A. Schwartz and R.O. Hynes, Cell, 31, 253–262
      (1982).

FIBRONECTIN: ROLE IN CELL SURFACE INTERACTIONS

John R. Couchman and Anne Woods

Biosciences Division
Unilever Research, Colworth Laboratory
Sharnbrook, Bedford MK44 1LQ, U.K.

BACKGROUND

Since the early 1970's, interest in fibronectin has increased steadily, and there is now a massive literature on the structure and functional attributes of this complex molecule (see reviews 1-3). Although its absence or depletion from the surface of transformed cells first brought fibronectin to the attention of those interested in tumour formation and invasion,[4-5] the fact that it has significant roles to play in the adhesion and migration of normal cells has widened the field to include connective tissue, developmental biology and more recently molecular biology. Fibronectin is now the most studied of all connective tissue components and two major attributes make this glycoprotein of particular interest. Firstly, the glycoprotein has the capacity to bind other connective tissue components and cell surfaces through a series of domains joined by more flexible regions[1] (Fig. 1). The implication of this is that fibronectin is ideally suited for a role in mediating cell attachment (whether eukaryotic or prokaryotic) to collagenous extracellular matrices. Secondly, the molecule is widespread in vivo, deposited in connective tissues and basement membranes and present in a soluble form in many body fluids including plasma, cerebrospinal and amniotic fluids.[1-3] It is also synthesised by a wide range of cell types[1,3] (Table 1) indicating a basic set of important functions of this molecule in cell and tissue homeostasis. These attributes will be further elaborated in succeeding sections.

21

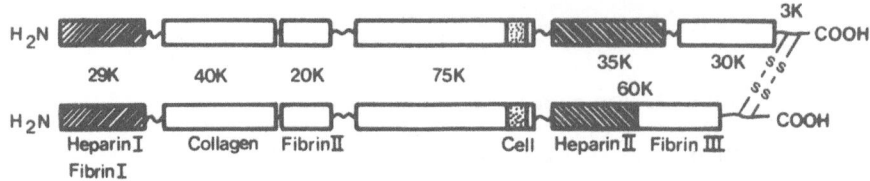

Fig. 1.   Schematic representation of plasma fibronectin. The
          disulphide-bonded dimer consists of a series of domains
          connected by flexible regions. Approximate molecular
          weights of the domains are shown (after Yamada[1]).

FIBRONECTIN STRUCTURE

     Plasma fibronectin exists as a dimer (Fig. 1) and has been
the most studied of the various forms. With the selective use of
proteases, affinity chromatographic techniques and cell attachment
assays, a map of the domains has been elucidated (Fig. 1). Of
these domains, one binds to collagens I to V,[1,6,7] although having
higher affinity for gelatin, two (or three)[8] bind to heparin with
differing affinities and three bind to fibrin.[1,2] Clearly this
gives fibronectin the ability to interact with a variety of other
extracellular macromolecules. In addition, fibronectin can bind
directly to a large number of vertebrate cell types and promote
their adhesion, leading in some cases to spreading and migration.
[3,9,10] Some bacteria can also bind fibronectin[11] through the
amino-terminal domain, and this may facilitate pathogenesis. The
primary structure of nearly the whole length of plasma fibronectin
has now been reported and for much of the molecule some information regarding its secondary structure is available (Fig. 2).
There is quite clearly the possibility that repeating homologous
structures within each domain arose by gene duplication, and
phylogenetic studies have shown that fibronectin-like molecules
are an ancient group, being found in invertebrates and even
sponges.[12] This data persuasively argues that the functions of
fibronectin are fundamental and a correlate of multicellularity.

DIFFERENCES BETWEEN FIBRONECTINS

     Although plasma fibronectin exists as a dimer, molecules of
fibronectin may interact to form insoluble multimers.[1] Differences in the structure between soluble fibronectin and the
insoluble connective tissue form have been sought and recent

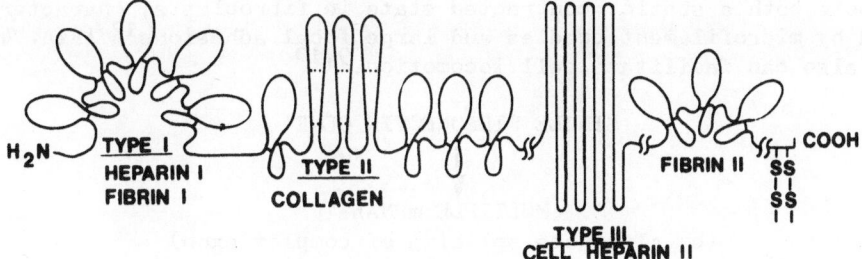

Fig. 2.   A monomer of fibronectin showing the different types of
          structural homology (Types I-III) found within its
          domains. The five repeats of Type I homology within the
          NH -terminal domain for example clearly indicate that
          they may have arisen by gene duplication events (after
          Yamada[1] and references therein).

evidence shows that some important variations in primary structure
can be identified. Although it appears that fibronectins are
derived from a single gene, there can be alternate splicing within
a complex exon to derive multiple forms of fibronectin message
differing in the presence or deletion of fairly short lengths of
mRNA.[13,14] The sites of these optional insertions of the mRNA are
in the carboxy-terminal one-third of the processed polypeptide.
It already appears that due to these options, there can be at
least six different fibronectin polypeptides which may go a long
way to defining the differences previously detected between the
two chains comprising each dimer of plasma fibronectin, and those
between the plasma and connective tissue forms of fibronectin.
[13-15] Presumably, one or more of the additional inserts which
code for the connective tissue form of fibronectin contains a
site which allows polymerisation to occur and thus make this form
more insoluble. Additionally, there are post-translational modifications including glycosylation[1] (Fig. 3) which may also differ
when plasma and connective tissue fibronectins from the same
species are compared. Most carbohydrate is, however, located in
the gelatin-binding domain of the molecule.[16]

ROLES OF FIBRONECTIN: IN VITRO STUDIES

     Much of our current idea of the function of fibronectin comes
from tissue culture experience. When adsorbed to a planar substratum, fibronectin can promote the attachment and spreading of
a wide variety of cell types including fibroblasts, endothelial
and epithelial cells.[1,17-19] Indeed this can occur in the absence
of serum or endogenously derived proteins.[17] However, an important consideration here is the type of adhesion that can be
fibronectin-promoted. It has now been shown that fibronectin can

promote both a static, contracted state in fibroblasts, characterised by microfilament bundles and large focal adhesions[17] (Fig. 4a)
and also can facilitate cell locomotion.[9,10]

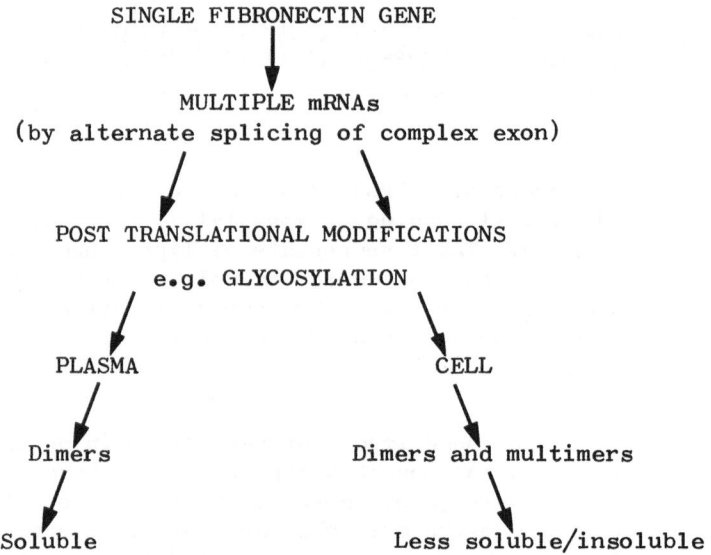

Fig. 3.   Scheme indicating differences in protein structure,
          glycosylation and properties between plasma and 'cell'
          derived fibronectins.

This latter behaviour requires a rather different cytoskeletal
organisation where microfilament-based activity leads to directed
protrusion and prominent microfilament bundles may not be formed
[20,21] (Fig. 4b). These contrasting types of adhesion clearly
require a different type of interaction with cell surfaces and
this is discussed in more detail below. However, a common feature
of the adhesive activity of fibronectin is that it leads to intracellular organisation and not merely a flattening of the cell, as
seen for example when fibroblasts are plated on to substrate-
adsorbed lectins or platelet factor-4.[22]

     Since plasma fibronectin can bind a variety of cellular and
extracellular components, and macrophages have been shown to
internalise fibronectin-coated particles,[23] plasma fibronectin has
also been postulated to act as an opsonin to enable debris clearance by the reticuloendothelial system.[24]

POSSIBLE ROLES OF FIBRONECTIN IN VIVO

     Several roles for fibronectin have been suggested (Table 2).
The possible involvement of plasma fibronectin in opsonisation has
just been discussed. Some of the other roles have a wound healing

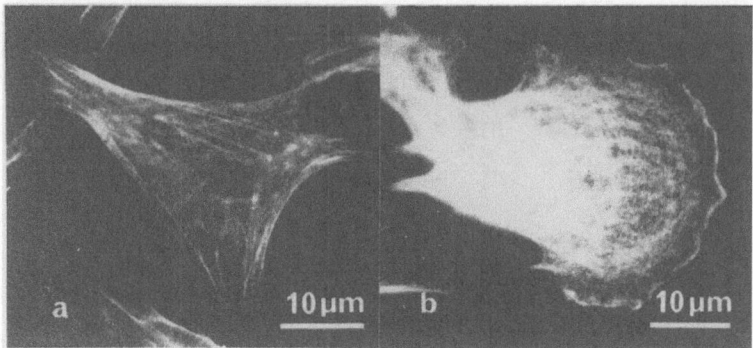

Fig. 4. Microfilament distributions in a static (a) and locomotory fibroblast (b) displayed by indirect immunofluorescence microscopy with antibodies to smooth muscle
myosin. Note the differences in cell shape as well as
the presence (a) or absence (b) of microfilament bundles.

and trauma connotation. Fibronectin can be covalently crosslinked to itself, to fibrin(ogen) and collagen[25] and so may have a
hemostatic role where tissue injury leads to vascular damage.
[1,2,25] Indeed severe trauma can lead to a dramatic decrease in
plasma fibronectin levels presumably resulting from fibronectin
deposition in thrombi and binding to exposed collagenous matrices
at the site of injury.[26] The levels of fibronectin at sites of
superficial skin damage do increase markedly as seen by immunofluorescence (Fig. 5) both in the fibrin containing clot and by
adsorption on to the collagen-rich dermis. Most of this fibronectin is of plasma origin and may form a provisional matrix for
the cellular processes of re-epithelialisation, wound closure and
new extracellular matrix deposition.[27] At later stages it has
also been shown that cellular fibronectin levels increase, particularly from the re-organizing vascular network.[28]

The functions of connective tissue fibronectin are uncertain
but there is substantial evidence from studies of embryogenesis
and tissue differentiation that large amounts of fibronectin are
associated with immature or undifferentiated tissues, often
decreasing with maturation.[29,30] In addition, many pathways for
migration in early embryonic development have fibronectin-rich
carpets over which movement takes place.[31-33] Such evidence for a
role in migration is by itself circumstantial, not least because
other components such as proteoglycans may also be plentiful, but
two recent reports have contributed some direct evidence for a
role of fibronectin in cell migration.[34,35] These pathways of
fibronectin for cell migration may also be linearly oriented so
that some element of contact guidance[36] for directionality of the
migrating cells may be important. Although several studies in
vitro have also implicated fibronectin with a role in cell

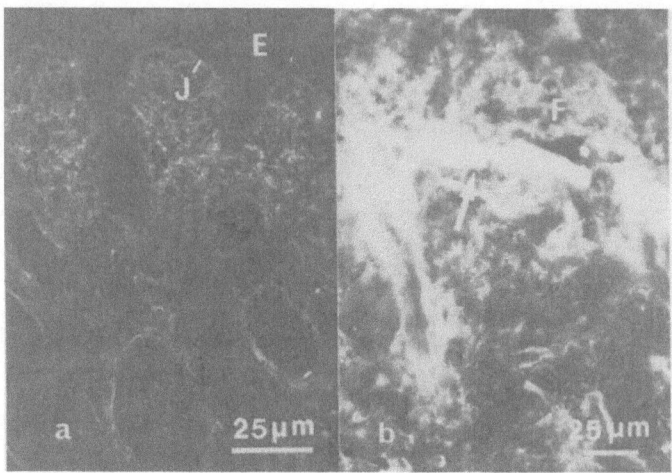

Fig. 5.   Fibronectin distributions in sections of normal rat skin
          (a) and after removal of an epidermal biopsy (b).
          Heavy staining is associated with the exposed dermal
          surface (arrow) and fibrin clot (F). Epidermis-E,
          dermo-epidermal junction-J.

migration,[9,10,37] this property may well not be exclusive for
fibronectin. The basement membrane glycoprotein laminin for
example has been shown to be a potent promoter of neurite outgrowth.[38] Above and beyond any role for fibronectin in the locomotory process of cell migration, it is apparent from a consideration of the structure of fibronectin that it may provide the
basis for more permanent cellular interactions with the extracellular matrix, whether for epithelial cell anchorage to basement
membranes or for mesenchymal cells enmeshed in a collagenous
milieu.

INTERACTIONS AT THE CELL SURFACE

    To understand how cell behaviour is guided by the extra-
cellular matrix, we clearly need insight into the mechanism by
which fibronectin and other connective tissue components interact
with the cell surface, leading to a triggering of cytoskeletal
organisation. Since many experiments have shown such a transmembrane interaction in tissue culture,[39,40] much effort has been
directed into the search for specific cell surface 'receptors'. A
few candidates such as a protein of 47,000 daltons[41] and glyco-
lipids[42] have been proposed, but recent developments have indicated
other cell surface components.[43,44] One problem has been that
individual fibronectin molecules appear to show only low affinity
binding to cells, which has rendered affinity chromatographic
techniques rather difficult. Indeed, earlier studies showed that
a critical density of fibronectin molecules was needed on the

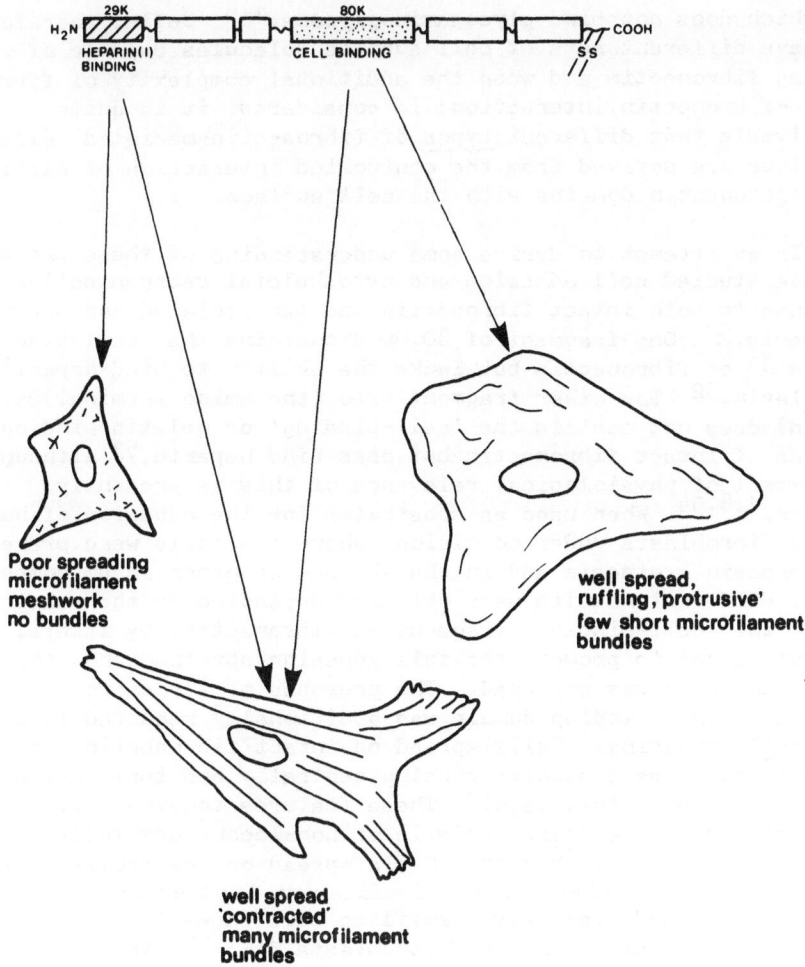

Fig. 6. Schematic summary of preliminary experiments on fibroblast spreading induced by the 29Kd and 80Kd domains of
human fibronectin either separately or combined.

substrate before induction of BHK cell spreading.[45] Additionally,
it seems highly plausible that more than one class of cell surface
'receptor' exists on some cell types. Various fibroblasts and
epithelial cells have been shown to carry on their surfaces a form
of heparan sulphate proteoglycan which has hydrophobic properties
and probably has one end inserted into the lipid of the cell membrane.[46-48] Not only that, but this cell surface molecule codis-
tributes closely with microfilament distributions in spread and
perturbed fibroblasts.[49] Fibronectin contains at least two
distinct sites to which heparin (and probably heparan sulphate)
can bind in addition to the cell binding domain which is distinct,

and which does not bind glycosaminoglycans.[1-8] Cells, therefore,
may have different sets of cell surface molecules capable of
binding fibronectin and when the additional complexity of fibro-
nectin-fibronectin interactions is considered, it is quite
conceivable that different types of fibronectin-mediated cellular
behaviour are derived from the controlled interaction of particular fibronectin domains with the cell surface.

In an attempt to derive some understanding of these processes
we have studied cell adhesion and cytoskeletal reorganisation in
response to both intact fibronectin and two isolated and purified
fragments.[50] One fragment of 80,000d contains the 'cell-binding
domain'[51] of fibronectin but lacks the ability to bind heparin
or gelatin.[52] The other fragment used (the amino terminal 29,000d
domain) does not contain the 'cell-binding' or gelatin-binding
domains of intact fibronectin but does bind heparin,[53] although
the extent of physiological relevance of this is presently
unclear.[1,8,54] When used as substrates for the adhesion of human
embryo fibroblasts under conditions where the cells were prevented
from protein synthesis and in the absence of other serum components, different results were obtained depending on the substrate
used. The 'cell-binding' fragment of fibronectin, by itself, was
not sufficient to promote the full adhesion obtained when the
intact molecule was employed. The presence of the amino-terminal
29,000d heparin-binding domain was additionally required to obtain
full cell spreading. Cells spread on intact fibronectin possess
large microfilament bundles running centrally and terminating in
focal adhesions (Figs. 4a,6). These features together with con-
cave cell edges are characteristic of non-locomotory fully
adherent fibroblasts.[10,21,22] Cells spread on the 'cell-binding
domain', however lacked large microfilament bundles or focal
adhesions and had protrusive, ruffling cell edges (Fig. 6)
more indicative of locomotory fibroblasts.[9,21,22] The cellular
response characteristic for intact fibronectin could, however, be
induced by a mixture of the two fragments ('cell' and 'heparin'
binding), but the amino-terminal 'heparin' binding domain alone was
a poor substrate for cell spreading (Fig. 6).

Two immediate conclusions can be drawn from these early
results. Firstly, promotion of fibroblast spreading does not
require the bivalency of dimeric fibronectin, and the cell binding
domain is sufficient. Secondly, two distinct signals need to be
given to spreading fibroblasts to initiate a full adhesion
characterised by microfilament bundles and focal adhesions.
Moreover these stimuli need not be covalently linked on the same
molecule. A most intriguing implication for the effects of fibronectin on cell behaviour is that the interaction of the cell-
binding domain alone with its receptor stimulates one type of
adhesion, whereas interaction with more than one type of domain
leads to a different morphology and behaviour. There is clearly

a great deal more to learn of the cell response to connective
tissue macromolecules and the control of transmembrane events
which follow. It seems likely that this complexity of cell surface interactions with extracellular matrix components is not
restricted to fibronectin. Although less well understood at the
present time it appears that the basement membrane glycoprotein
laminin[1,55] is also capable of multi-domain interactions with cell
surfaces.[56,57]

FIBRONECTIN, WOUND HEALING AND TISSUE REPAIR

Finally, we will speculate on the pluripotential roles of
fibronectin in response to tissue damage, such as a skin lesion.
We have already referred to a role for this molecule in hemostasis
and clot formation and it is likely that plasma fibronectin from
disrupted blood vessels binds to fibrin, an interaction stabilised
by the transglutaminase factor XIIIa.[25] By virtue of its interaction with many other connective tissue components, including
collagens I-V,[6-7] proteoglycans[8,54] and hyaluronic acid[58] it is
further quite likely that plasma fibronectin deposits in the
damaged tissues in large amounts (Fig. 5). In addition, platelets
contain fibronectin which may be released on activation at the
site of damage.[59] The fibronectin-rich wound bed then serves as a
temporary matrix into which cell migration can occur. We and
others have shown that fibroblast migration can be fibronectin-
mediated[9,10] and studies on major pathways of migration in
vertebrate embryos have shown them to be rich in fibronectin.[31,35]
It may further be that epithelial migration for epidermal repair
is fibronectin-mediated;[60,61] we have some preliminary data which
show that rat epidermal cell migration is inhibited by affinity
purified antibodies to fibronectin in vitro.[62] What type of cell
interactions with fibronectin promote the migratory responses?
This is presently unknown, but two interesting corollaries of our
cell adhesion experiments are worth noting. Firstly, the 'cell-
binding' domain of fibronectin has been reported to be chemotactic
to an extent which may be greater than the intact molecule.[63,64]
Furthermore, fibronectin lacking the amino-terminal domain may be
generated at the wound site since some proteases possibly released
there easily cleave between this domain from the remainder of the
molecule[65] (Fig. 7). It could be that the cleaved fibronectin
promotes a 'motile' phenotype for cell migration. At the same
time however, it is clear that the mesenchymal cells can form
'static' adhesions with the surrounding granulation tissue since
a recent report has elegantly shown fibroblasts with large microfilament bundles (typical of stable adhesion) interacting with
the fibronectin-rich matrix of a wound bed.[66] Wound closure may
indeed result from the co-operative contraction exerted by these
cells on their extracellular matrix.

Debris clearance from the wound bed may also have a fibro-

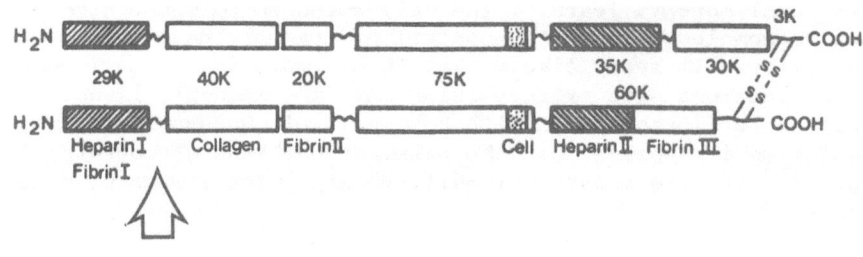

                                    PLASMIN

                                    LEUKOCYTE ELASTASE

                                    THROMBIN

                                    TRYPSIN

                                    SUBTILISIN

Fig. 7. Plasma fibronectin showing a number of enzymes which
         cleave between the NH2-terminal heparin binding domain
         and the remainder of the molecule. Some of these may be
         present at sites of tissue damage. (See text.)

nectin component since it has been shown that it can act as an
opsonin for particle uptake by phagocytes.[2,24] After cell migration into the wound is complete the scab has been lost and vascular reorganisation has occurred, the amount of fibronectin at the
wound site gradually diminishes[67] in a similar way to that seen in
skin development, where the extent of fibronectin staining diminishes in the mature tissues.[30] Thus wound repair mimics development to some extent. It can be seen that fibronectin may have
many different roles to fulfil in the repair of a simple wound,
principally because of its potential to bind many other extracellular matrix and plasma components as well as the surface of
many cell types. The responses are however quite different,
whether it be migration (and possibly chemotaxis), phagocytosis
or wound contraction. The common feature of all these is of
course the transmembrane stimulation of the microfilamentous
compartment of the cytoskeleton following the interaction of
fibronectin with cell surface 'receptors'. It seems highly likely
that this ligand may bind to more than one species of 'receptor'
in these cell types and even on one cell type as indicated by our
adhesion studies. We still await confirmation of the identity of
the 'receptor(s)' which interact with the 'cell-binding' domain of
fibronectin. The possibility that heparin binding domains of
fibronectin interact with a membrane-intercalated heparan sulphate
proteoglycan also is apparent[48,49,68-70] but has yet to be
directly demonstrated.

In conclusion, we are left in no doubt that fibronectin interactions with cell surfaces are complex and can lead to a variety of cellular responses but appear also to be an important part of embryogenesis, tissue repair and the function of the reticulo endothelial system.

Table 1. Cell Types producing Fibronectin in vitro

Fibroblasts

Endothelial Cells

Myoblasts

Epithelial Cells

   e.g. keratinocytes
       amniotic cells
       hepatocytes

Endodermal, including teratocarcinoma

Glial Cells?

Chondrocytes

Platelets

Neutrophils, Macrophages and Mast Cells

Not produced by Lymphocytes or Erythrocytes

Table 2. Possible Roles of Fibronectin <u>in</u> <u>vivo</u>

Extracellular Matrix Formation

through interactions with other components -

e.g. collagens
      proteoglycans
      fibrin

Adhesion of Cells to Extracellular Matrices

e.g. connective tissue
      basement membrane

Migration

e.g. embryogenesis
      tissue repair

Hemostasis and Thrombosis

Phagocytosis

ACKNOWLEDGEMENT

We thank Mrs. Gill Pattullo for her skilled assistance in
the preparation of this manuscript.

REFERENCES

1. K.M. Yamada, Cell surface interactions with extracellular
       materials, Ann. Rev. Biochem. 52:761 (1983).
2. D.F. Mosher, Physiology of fibronectin, Ann. Rev. Med. 35:561
       (1984).
3. R.O. Hynes, Fibronectin and its relation to cellular structure
       and behavior, in "Cell Biology of Extracellular Matrix",
       E.D. Hay, ed., Plenum Press, New York (1981).
4. A. Vaheri and D.F. Mosher, High molecular weight cell surface-
       associated glycoprotein (fibronectin) lost in malignant
       transformation, Biochim. Biophys. Acta 516:1 (1978).
5. R.O. Hynes and K.M. Yamada, Fibronectins: Multi-functional
       modular glycoproteins, J. Cell Biol. 95:369 (1982).
6. E. Engvall, E. Ruoslahti, and E.J. Miller, Affinity of fibro-
       nectin to collagens of different genetic types and to
       fibrinogen, J. Exp. Med. 147:1584 (1978).
7. F. Jilek and H. Hörmann, Cold-insoluble globulin (fibronectin)
       IV. Affinity to soluble collagen of various types, Hoppe-
       Seyler's Z. Physiol. Chem. 359:247 (1978).
8. L.I. Gold, B. Frangione, and E. Pearlstein, Biochemical and
       immunological characterization of three binding sites on
       human plasma fibronectin with different affinities for
       heparin, Biochemistry 22:4113 (1983).
9. I.V. Ali and R.O. Hynes, Effects of LETS glycoprotein on cell
       motility, Cell 14:439 (1978).
10. J.R. Couchman, D.A. Rees, M.R. Green, and C.G. Smith, Fibro-
       nectin has a dual role in locomotion and anchorage of
       primary chick fibroblasts and can promote entry into the
       division cycle, J. Cell Biol. 93:402 (1982).
11. G. Fröman, L. Switalski, A. Faris, T. Wadstrom, and M. Höök,
       Binding of E. coli to fibronectin - A mechanism of tissue
       adherence, J. Biol. Chem. In Press (1985).
12. S.K. Akiyama and M.D. Johnson, Fibronectin in evolution:
       Presence in invertebrates and isolation from Microciona
       prolifera, Comp. Biochem. Physiol. B. 76:687 (1983).
13. J.E. Schwarzbauer, J.W. Tamkun, I.R. Lemischka, and R.O. Hynes,
       Three different fibronectin mRNAs arise by alternative
       splicing within the coding region, Cell 35:421 (1983).
14. A.R. Kornblihtt, K. Vibe-Pedersen, and F.E. Baralle, Human
       fibronectin: molecular cloning evidence for two mRNA
       species differing by an internal segment coding for a
       structural domain, EMBO J. 3:221 (1984).
15. J.W. Tamkun, J.E. Schwarzbauer, and R.O. Hynes, A single rat
       fibronectin gene generates three different mRNAs by
       alternative splicing of a complex exon, Proc. Natl. Acad.
       Sci. U.S.A. 81:5140 (1984).
16. K. Sekiguchi and S. Hakomori, Functional domain structure of
       fibronectin, Proc. Natl. Acad. Sci. U.S.A. 77:2661 (1980).

17. J.R. Couchman, M. Höök, D.A. Rees, and R. Timpl, Adhesion, growth and matrix production by fibroblasts on laminin substrates, J. Cell Biol. 96:177 (1983).

18. D.M. Scott, J.C. Murray, and M.J. Barnes, Investigation of the attachment of bovine corneal endothelial cells to collagens and other components of the subendothelium, Exp. Cell Res. 144:472 (1983).

19. M. Höök, K. Rubin, A. Oldberg, B. Öbrink, and A. Vaheri, Cold-insoluble globulin mediates the adhesion of rat liver cells to plastic petri dishes, Biochem. Biophys. Res. Commun. 79:726 (1977).

20. R.A. Badley, J.R. Couchman, and D.A. Rees, Comparison of the cell cytoskeleton in migratory and stationary chick fibroblasts, J. Muscle Res. Cell Motility 1:5 (1980).

21. J.R. Couchman, M. Lenn, and D.A. Rees, Coupling of cytoskeleton functions for fibroblast locomotion, Eur. J. Cell Biol. In Press (1985).

22. R.J. Beyth and L.A. Culp, Complementary adhesive responses of human skin fibroblasts to the cell-binding domain of fibronectin and the heparan sulfate-binding protein, platelet factor-4, Exp. Cell Res. 155:537 (1984).

23. J.E. Doran, A.R. Mansberger, and A.C. Pease, Cold-insoluble globulin-enhanced phagocytosis of gelatinized targets by macrophage monolayers: A model system, J. Retic. Soc. 27:471 (1980).

24. F.A. Blumenstock, T.M. Saba, P. Weber, and R. Laffin, Biochemical and immunological characterization of human opsonic α_2 SB glycoprotein: Its identity with cold-insoluble globulin, J. Biol. Chem. 253:4287 (1978).

25. F. Jilek and H. Hörmann, Cold-insoluble globulin. III. Cyanogen bromide and plasminolysis fragments containing a label introduced by transamidation, Hoppe-Seyler's Z. Physiol. Chem. 358:1165 (1977).

26. A.B. Robbins, J.E. Doran, A.C. Reese, A.R. Mansberger, Cold-insoluble globulin levels in operative trauma: serum depletion, wound sequestration and biological activity: an experimental and clinical study, Am. Surg. 46:663 (1980).

27. R.A.F. Clark, H.J. Winn, H.F. Dvorak, and R.P. Colvin, Fibronectin beneath reepithelializing epidermis in vivo; sources and significance, J. Invest. Dermatol. 80:265 (1983)

28. R.A.F. Clark, P. DellaPelle, E. Manseau, J.M. Lanigan, H.F. Dvorak, and R.B. Colvin, Blood vessel fibronectin increases in conjunction with endothelial cell proliferation and capillary ingrowth during wound healing, J. Invest. Dermatol. 79:269 (1982).

29. J.R. Couchman, W.T. Gibson, D. Thom, A.C. Weaver, D.A. Rees, and W.E. Parish, Fibronectin distribution in epithelial and associated tissues of the rat. Arch. Dermatol. Res. 266:295 (1979).

30. W.T. Gibson, J.R. Couchman, and A.C. Weaver, Fibronectin distribution during the development of fetal rat skin, J. Invest. Dermatol. 81:480 (1983).

31. D.R. Critchley, M.A. England, J. Wakely, and R.O. Hynes, Distribution of fibronectin in the ectoderm of gastrulating embryos, Nature, Lond. 280:498 (1979).

32. B.W. Mayer, E.D. Hay, and R.O. Hynes, Immunocytochemical localization of fibronectin in embryonic chick trunk and area vasculosa, Dev. Biol. 82:267 (1981).

33. J. Heasman, R.O. Hynes, A.P. Swan, V. Thomas, and C.C. Wylie, Primordial germ cells of Xenopus embryos: the role of fibronectin in their adhesion during migration, Cell 27: 437 (1981).

34. J.C. Boucaut, T. Darnibere, H. Boulebache, and J.P. Thiery, Prevention of gastrulation but not neuralation by antibodies to fibronectin in amphibian embryos, Nature, Lond. 307:364 (1984).

35. J.P. Thiery, J.-L. Duband, A. Delouvee, G. Tucker, H. Aoyama, T.J. Poole, and K.M. Yamada. Ontogeny of the peripheral nervous system, J. Embryol. exp. Morphol. 82:35 (1984).

36. G.A. Dunn, Contact guidance of cultured tissue cells: a survey of potentially relevant properties of the substratum, in "Cell Behaviour", R. Bellairs, A.S.G. Curtis, G.A. Dunn, eds., Cambridge University Press, Cambridge (1982).

37. D.C. Turner, J. Lawton, P. Dollenmeier, R. Ehrismann, and M. Chiquet, Guidance of myogenic cell migration by oriented deposits of fibronectin, Dev. Biol. 95:497 (1983).

38. A. Baron-Von Evercooren, H.K. Kleinman, S. Ohno, P. Marangos, I.P. Schwartz, and M.E. Dubois-Dalcq, Nerve growth factor, laminin and fibronectin promote neurite outgrowth in human fetal sensory ganglion cultures, J. Neurosci. Res. 8:179 (1982).

39. R.O. Hynes and A.T. Destree, Relationships between fibronectin (LETS protein) and actin, Cell 15:875 (1977).

40. V.-P. Lehto, T. Vartio, and I. Virtanen, Fibronectin remains in the cytoskeletal preparations of cultured human fibroblasts, Cell Biol. Int. Rep. 5:417 (1981).

41. J.D. Aplin, R.C. Hughes, C.L. Jaffe, and N. Sharon, Reversible cross-linking of cellular components of adherent fibroblasts to fibronectin and lectin-coated substrata, Exp. Cell Res. 134:488 (1981).

42. H.K. Kleinman, G.R. Martin, and P.H. Fishman, Ganglioside inhibition of fibronectin-mediated cell adhesion to collagen, Proc. Natl. Acad. Sci. U.S.A. 76:3367 (1979).

43. P.J. Brown and R.L. Juliano, Admodulin: A cell surface glycoprotein specifically involved in fibronectin-mediated adhesion, J. Cell Biol. 99:161a (1984).

44. T. Hasegawa, E. Hasegawa, W.-T. Chen, and K.M. Yamada,
    Characterization of a membrane glycoprotein complex
    implicated in cell adhesion to fibronectin, J. Cell Biol.
    99:165a (1984).
45. R.C. Hughes, S.D.J. Pena, J. Clark, and R.R. Dourmashkin,
    Molecular requirements for the adhesion and spreading of
    hamster fibroblasts, Exp. Cell Res. 121:307 (1979).
46. L. Kjellén, I. Pettersson, and M. Höök, Cell-surface heparan
    sulfate: an intercalated membrane proteoglycan, Proc. Natl.
    Acad. Sci. U.S.A. 78:5371 (1981).
47. A.C. Rapraeger and M. Bernfield, Heparan sulfate proteoglycans
    from mouse mammary epithelial cells, J. Biol. Chem. 258:
    3632 (1983).
48. L.Å. Fransson, I. Carlstedt, L. Cöster, and A. Malmstrom,
    Structure and function of cell-surface associated proteo-
    heparan sulphate, Eur. J. Cell Biol. 1:18 (1983).
49. A. Woods, M. Höök, L. Kjellén, C.G. Smith, and D.A. Rees,
    Relationship of heparan sulfate proteoglycans to the
    cytoskeleton and extracellular matrix of cultured fibro-
    blasts, J. Cell Biol. 99:1743 (1984).
50. A. Woods, J.R. Couchman, M. Höök, and S. Johansson, Adhesion
    and cytoskeletal organisation of fibroblasts in response
    to fibronectin peptides. In Preparation.
51. M.D. Pierschbacher and E. Ruoslahti, Cell attachment activity
    of fibronectin can be duplicated by small synthetic
    fragments of the molecule, Nature, Lond. 309:30 (1984).
52. S. Johansson, Demonstration of high affinity fibronectin-
    receptors on rat hepatocytes in suspension, J. Biol. Chem.
    In Press (1985).
53. A. Garcia-Pardo, E. Pearlstein, and B. Frangione, Primary
    structure of human plasma fibronectin. The 29,000-dalton
    NH$_2$-terminal domain, J. Biol. Chem. 258:12670 (1983).
54. K. Sekiguchi, S. Hakomori, M. Funahashi, I. Matsumoto, and
    N. Seno, Binding of fibronectin and its proteolytic
    fragments to glycosaminoglycans, J. Biol. Chem. 258:14359
    (1983).
55. R. Timpl, H. Rohde, P.G. Robey, S.I. Rennard, J.M. Foidart,
    and G.R. Martin, Laminin - A glycoprotein from basement
    membranes, J. Biol. Chem. 254:9933 (1979).
56. D. Edgar, R. Timpl, and H. Thoenen, The heparin-binding
    domain of laminin is responsible for its effects on
    neurite outgrowth and neuronal survival, EMBO J. 3:1463
    (1984).
57. R. Timpl, S. Johansson, V. van Delden, I. Oberbaumer, and
    M. Höök, Characterization of protease-resistant fragments
    of laminin mediating attachment and spreading of rat
    hepatocytes, J. Biol. Chem. 258:8922 (1983).

58. J. Jilek and H. Hörmann, Fibronectin (cold-insoluble globulin).
    VI. Influence of heparin and hyaluronic acid on the
    binding of native collagen, Hoppe-Seyler's Z. Physiol.
    Chem. 360:597 (1979).
59. M.H. Ginsberg, R.G. Painter, J. Forsyth, C. Birdwell, and
    E.F. Plow, Thrombin increases expression of fibronectin
    antigen on the platelet surface, Proc. Natl. Acad. Sci.
    U.S.A. 77:1049 (1980).
60. W.T. Gibson, J.R. Couchman, R.A. Badley, H.J. Saunders, and
    C.G. Smith, Fibronectin in cultured rat keratinocytes:
    distribution, synthesis, and relationship to cytoskeletal
    proteins, Eur. J. Cell Biol. 30:205 (1983).
61. D.J. Donaldson and J.T. Mahan, Fibrinogen and fibronectin as
    substrates for epidermal cell migration during wound
    closure, J. Cell Sci. 62:117 (1983).
62. J.R. Couchman and S. Blencowe, Adhesion and cell surface
    relationships during fibroblast and epithelial migration
    in vitro, in "Cell Traffic in the Developing and Adult
    Organism", G. Haemmerli and P. Sträuli, eds., Karger,
    Basel. In Press (1985).
63. A.E. Postlethwaite, J. Keski-Oja, G. Balian, and A.H. Kang,
    Induction of fibroblast chemotaxis by fibronectin.
    Localization of the chemotactic region to a 140,000-
    molecular weight non-gelatin-binding fragment, J. Exp.
    Med. 15:194 (1981).
64. H.E.J. Seppa, K.M. Yamada, S.T. Seppa, M.H. Silver,
    H.K. Kleinman, and E. Schiffman, The cell binding fragment
    of fibronectin is chemotactic for fibroblasts, Cell Biol.
    Int. Rep. 5:813 (1981).
65. M.B. Furie and D.B. Rifkin, Proteolytically derived frag-
    ments of human plasma fibronectin and their localization
    within the intact molecule, J. Biol. Chem. 255:3134 (1980).
66. I.I. Singer, D.W. Kawka, D.M. Kazazis, and R.A.F. Clark, In
    vivo co-distribution of fibronectin and actin fibers in
    granulation tissue: immunofluorescence and electron
    microscope studies of the fibronexus at the myofibroblast
    surface, J. Cell Biol. 98:2091 (1984).
67. M. Kurkinen, A. Vaheri, P.J. Roberts, and S. Steinman,
    Sequential appearance of fibronectin and collagen in
    experimental granulation tissue, Lab. Invest. 43:47 (1980).
68. M.W. Lark and L.A. Culp, Multiple classes of heparan sulfate
    proteoglycans from fibroblast substratum adhesion sites,
    J. Biol. Chem. 259:6773 (1984).
69. M. Höök, L. Kjellén, S. Johansson, and J. Robinson, Cell sur-
    face glycosaminoglycans, Ann. Rev. Biochem. 53:847 (1984).
70. A. Woods, J.R. Couchman, and M. Höök, Heparan sulfate proteo-
    glycans of rat embryo fibroblasts. A hydrophobic form may
    link cytoskeleton and matrix components, J. Biol. Chem.
    Submitted.

CYTOSKELETAL AND CYTOCONTRACTILE FEATURES OF MYOFIBROBLASTS

Giulio Gabbiani

Department of Pathology
University of Geneva
1211 Geneva 4, Switzerland

INTRODUCTION

Two phenomena play an essential role for the closing of an open wound: one is formation and contraction of granulation tissue, and the second is epithelialization, i.e. movement and replication of epithelial cells over the wounded area. Similar to placenta during pregnancy, granulation tissue is a new and temporary organ which disappears as soon as the wound is closed by epithelialization. The main functions of granulation tissue are: 1) synthesis of new connective tissue, and 2) production of a contractile movement which brings together the margins of the wound. Old experiments by Carrel [1] had shown that this contractile force is produced within the granulation tissue itself. We have studied the morphologic, functional and pharmacological characteristics of fibroblasts under normal conditions and during wound healing or fibrocontractive diseases. Our results indicate that during wound healing and fibrocontractive diseases, fibroblasts assume several characteristics of smooth muscle cells. These modified fibroblasts or myofibroblasts probably play the key role in granulation tissue contraction or in pathological connective tissue retractions. We shall now review the characteristics of normal fibroblasts and of myofibroblasts.

THE NORMAL FIBROBLAST

The fibroblast [2] was first identified by means of light microscopy on the base of its shape and its relationship with the extracellular substance. The use of electron microscopy has allowed a better definition of the cytological characteristics of fibroblasts. The nucleus is generally large and contains one or more nucleolo. The most prominent cytoplasmic organelle is rough endoplasmic

reticulum which consists of a series of interconnected sack-like
or tubular structures present throughout the cytoplasm. The
content of these cisterns is relatively dense and sometimes finely
filamentous[3,4]; ribosomes form large aggregates on the membrane[2,5].
The Golgi apparatus is generally prominent and has no particular
location[2]. Abundant mitochondria are present throughout the cytoplasm. Only a few cytoplasmic microfilaments (40-70 Å in diameter)
and intermediate filaments (100 Å in diameter) may be seen in fibroblasts of adult animals or humans particularly close to the plasmalemma[4]. In normal tissues of adult animals, there are no contacts
between fibroblasts. However, contacts can be seen between
embryonic and fetal fibroblasts as well as between fibroblasts of
newborn animals[6,8]. These contacts most commonly take the form of
tight junctions.

THE MYOFIBROBLAST

    In granulation tissue of normal wounds or in pathologic
connective tissue during fibrocontractive diseases (e.g. Dupuytren's
nodule), fibroblasts assume several new features.

Morphology

    A fibrillar system develops within the cytoplasm[9]; not the few
fibrils seen in normal fibroblasts, but bundles of parallel fibrils
resembling those of smooth muscle cells. Individual fibrils measure
40-80 Å in diameter (more rarely 100-120 Å) and are usually arranged
parallel to the long axis of the cell. Many electron-opaque areas
are scattered among the bundles or located beneath the plasmalemma.
These are similar to the attachment sites of smooth muscle.
Although the fibrillar structures often occupy a large portion of
the cell, the remaining cytoplasm contains packed cisterns of rough
endoplasmic reticulum typical of normal fibroblasts. The nuclei
show multiple indentations or deep folds, an appearance quite unlike
that of normal fibroblasts (or other cells in the same granulation
tissue such as macrophages or mast cells). There are numerous
intercellular connections between granulation tissue fibroblasts.
Their structure identifies them as gap junctions[10]. In addition,
part of the cell surface is often covered by a well-defined layer of
material having the structural features of a basal lamina and
generally separated from the cell membrane by a translucent layer.
Where it is covered by a basal lamina, the cell often shows dense
zones in the fibrillar bundles immediately beneath the surface membrane. The resulting complex is reminiscent of hemidesmosomes which
bind endothelial cells, pericytes, and smooth muscle cells to their
basal laminae.

Pharmacology

    Strips of granulation tissue from animals or humans placed in a pharmacological bath behave like smooth muscle in that they are contracted or relaxed by substances that contract or relax smooth muscle [11-13]. Among the substances most active in inducing contractions are: 5-hydroxytriptamine (5-HT or serotonin), angiotensin, vasopressin, norepinephrine, bradykinin, epinephrine, and prostaglandin F_{1a}.

Chemistry

The yield of actomyosin obtained by extraction from a croton oil-induced granuloma pouch (4.0 mg of actomyosin per gram wet weight of pouch tissue) is comparable to that obtained with identically prepared extracts of pregnant rat uteri (3.5 mg/g wet weight)[11]. The calcium-activated adenosine-triphosphatase activity of these extracts is similar, splitting approximately 10 nM of adenosine triphosphate per mg or protein per min.

The amount of actin in granulation tissue myofibroblasts, calculated as a percent of total protein by densitometric evaluation of monodimensional SDS-polyacrylamide gels [14], is intermediate between that of normal subcutaneous tissue and that of smooth muscle tissue, such as pregnant rat uterus [15].

An analysis of actin isoforms by means of bidimensional electrophoresis according to O'Farrel [16], shows that myofibroblasts contain β and γ-actin, similarly to most mesenchymal cells, but differently from smooth muscle cells which contain α, β and γ-isoforms [17]. This suggests that myofibroblasts derive from fibroblastic cells.

Immunology

Granulation tissue fibroblasts gradually develop intracellular neoantigens which are similar to those present in smooth muscle cells. Thus they fix anti-actin and antimyosin antibodies [10]. This gives further support to the possibility that a mechanism involving an interaction between actin and myosin is implicated in the contraction of granulation tissue. When granulation tissue disappears after the healing of a wound, no more fixation of anti-actin and antimyosin antibodies to granulation tissue fibroblasts is observed.

Myofibroblasts contain only vimentin as constituent of intermediate filaments in analogy with what is observed in most mesenchymal cells. The absence of desmin (the main constituent of intermediate filaments in muscular cells) from myofibroblasts is again in accordance with their fibroblastic origin.

Collagen Synthesis

In inflamed tissues, collagen is synthesized more rapidly and is present in a higher concentration than in normal tissues [18].

When the inflammatory reaction subsides, collagen is progressively
resorbed, and the repaired tissue returns to normal composition.
Collagen from acutely or chronically inflamed tissue is less
soluble than collagen from normal tissue. This corresponds to the
presence of granulation tissue collagen of crosslinks different from
those present in collagen of normal skin, but similar to those
present in collagen of embryonic skin [19,20]. Moreover, granulation
tissue induced in the rat by subcutaneous implantation of polyvinyl
sponges, contains a higher proportion of type III collagen than
normal skin [21]. Myofibroblasts are present while the tissue is
synthesizing type III collagen and disappears when normal type I
collagen with different stabilizing crosslinks is being synthesized [22]. Therefore it appears probable that myofibroblasts are,
at least in part, responsible for the synthesis of type III collagen.
The collagen in normal skin is almost totally of the classic type I,
the fibers of which possess a typical 640 Å periodicity, whereas in
granulation tissue it is composed of relatively few classic collagen
fibers, some fibers without periodicity, and a significant quantity
of finely filamentous material. It may be speculated that the small
filaments and the fibers without periodicity are composed mainly of
type III collagen. These fibers are probably analogous to those
generally referred to as reticulin. It is worth noting that type
III collagen is present in tissues that need a certain plasticity,
such as embryonic skin, normal smooth muscle and granulation tissue.
An increased amount of type III collagen has been found in nodules
of Dupuytren's disease compared with normal plamar aponeurosis [23].
Cultivated fibroblasts obtained from chronically inflamed human
gengiva produce increased amounts of type III collagen when compared
to fibroblasts obtained from non-inflamed tissues [24].

Tissue Culture

    Myofibroblasts have been cultured from normal granulation
tissue [25], as well as from fibromatotic tissues such as Peyronie's
disease [26], or the Dupuytren's nodule [27]. Myofibroblasts cultured
from Dupuytren's nodules display in vitro biological properties that
are intermediate between those expressed by normal fibroblasts and
sarcoma cells or cells from the nodule transformed with SV-40 virus.
Thus they represent an interesting in vitro model of partially
transformed human cells. This behaviour is not evolutive and
justifies the classification of Dupuytren's disease among the benign
mesenchymal tumours. The production of high level of plasminogen
activator by myofibroblasts cultured from Dupuytren's nodules [27]
explains at least in part the local reactive pathology and could act
as mitogenic stimulus for the proliferation of the nodule itself.
Cultures derived from the apparently normal palmar aponeurosis show
some but not all the abnormal growth properties of cells from
nodules. This may help to explain the onset of local recurrences.
These results support the view that Dupuytren's disease is not
strictly local and limited to the nodules, but affects at least

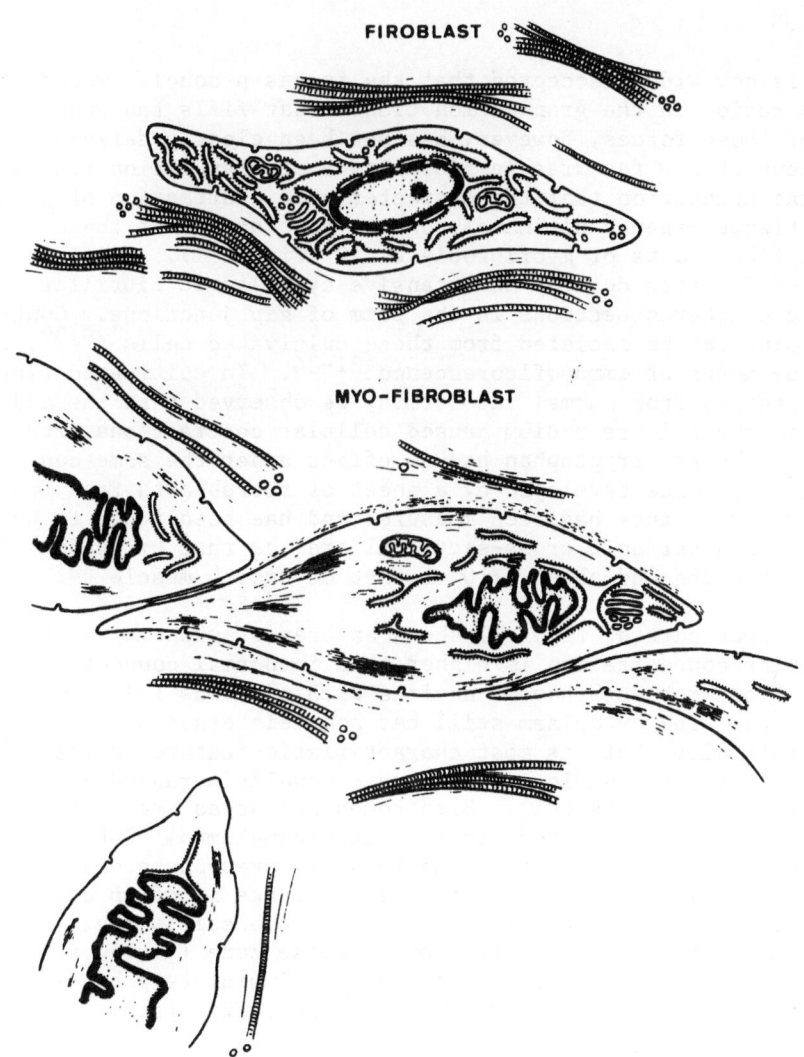

FIROBLAST

MYO-FIBROBLAST

Fig. 1. Scheme comparing the characteristics of fibroblasts. The upper part of the figure shows a typical fibroblast with a smooth contour of the nucleus which contains a nucleolus. The cytoplasm contains cisternae of rough endoplasmic reticulum, mitochondria, a Golgi apparatus, and peripheral vesicles, but only few intracytoplasmic fibrils.

partially the whole aponeurosis. Dupuytren's nodules could be considered as a model of tumour progression in a benign situation.

CONCLUSION

It is now widely accepted that the forces producing wound contraction reside in the granulation tissue that fills the wound. The nature of these forces, however, has not been clearly defined. The development of new features in fibroblasts of granulation tissue has led to the suggestion that the characteristic contraction of granulation tissue depends ultimately on the contraction of these modified fibroblasts or myofibroblasts [28] (Figure 1). Fibroblasts cultivated in vitro develop an extensive cytoplasmic fibrillar system, and interconnections in the form of gap junctions. Contractile proteins can be isolated from these cultivated cells [29,39], or stained by means of immunofluorescence [31-34]. In cultivated fibroblasts obtained from normal rat dermis, we observed that the addition of 5-HT to the culture medium caused cellular contractions within 15-20 min, whereas tryptophan had no effect under the same conditions [11]. The force developed by a sheet of fibroblasts free from the substratum in culture has been measured and has been found to be about the same per unit cross-sectional area as that of granulation tissue of a wound and about 1/10 of that of smooth muscle [35].

The lower part of Figure 1 shows an area of granulation tissue. The cellular concentration is higher than in normal connective tissue. Myofibroblasts have a nucleas with numerous folds and indentations. The cytoplasm still has some cisternae of rough endoplastic reticulum, but its most characteristic feature is the presence of massive bundles of filaments usually arranged parallel to the long axis of the cell. Electron-dense areas are scattered among the bundles or located beneath the plasmalemma. Intercellular connections in the form of gap junctions are present between fibroblasts. In addition, a part of the cell surface is often covered by a well defined layer of material similar to a basal lamina. In such regions, the cell commonly shows a dense zone (giving a hemidesmosome complex) in the fibrillar bundles immediately beneath the surface membrane (from Gabbiani, Meth. Achiev. Exp. Path. 9: 187, 1979).

Myofibroblasts have been described in man and animals during several pathologic situations. Some of these situations are inflammatory in nature and related to wound healing [36-48]. Some are connected to the so-called fibromatoses [50-61], and in some cases the pathogenesis is unknown, e.g. liver cirrhosis [62,63], or kidney fibrosis [64]. Finally, there are few reports of malignant myofibroblastic tumours [65-68].

It appears that in experimental animals or humans, the fibroblast is a very plastic cell which can adapt to different situations by changing morphological, biochemical and functional features. Thus, during wound healing there is a transformation of local fibroblasts and/or other less differentiated cells into myofibroblasts [28]. These acquire many properties typical of cultivated fibroblasts and in particular an important contractile apparatus which is probably responsible for the mechanism of granulation tissue contraction.

A relationship between intracytoplasmic filaments and cellular motion, development of tension, and intracytoplasmic movements or secretion has been proposed for a wide spectrum of cells ranging from monocellular organisms to those of mammalian tissues [69]. The development of a contractile filamentous apparatus probably takes place when cells of different embryological origin face situations that require the enhancement of certain characteristic functions, such as the ability to move about and to contract. In granulation tissue, the contraction of a single myofibroblast is then transmitted to other cells and to the stroma; this response is synchronized rather than individual.

ACKNOWLEDGEMENTS

We thank S. Karger, Basel, for granting the permission of reproducing Figure 1. Supported in part by the Swiss National Science Foundation, Grant Nr. 3.178-0.82 and by the E. and L. Schmidheiny Foundation.

REFERENCES

1. Carrel A., Cicatrisation of wounds. I. The relation between the size of a wound and the rate of its cicatrisation, J. Exp. Med. 24:429 (1916).
2. Ross R., The connective tissue fiber forming cell, in: "Treatise on Collagen", Vol. 2, Part A, G.N. Ramachandran, ed., Academic Press, New York (1968).
3. Movat H. Z. and Fernando N. V. P., The fine structure of connective tissue. I. The fibroblast, Exp. Mol. Pathol., 1:509 (1962).
4. Ross R., and Benditt E. P., Wound healing and collagen formation. Sequential changes in components of guinea pig skin wounds observed in the electron microscope, J. Biophys. Biochem. Cytol. 11:677 (1961).
5. Palade G. E., A small particulate component of the cytoplasm, in: "Rantiers in Cytology", L.P. Sandford, ed., Yale University Press, New Haven (1958).
6. Greenle T. K. and Ross R., The development of the rat flexor digital tendon. A fine structure study, J. Ultrastruct. Res. 18:354 (1967).
7. Ross R. and Greenle T.K., Electron microscopy: attachment sites between connective tissue cells, Science 153:997 (1966).

8. Trelstad R. L., Kang A. H., Igarashi S., and Gross J.,
    Isolation of two distinct collagens from chick cartilage,
    Biochemistry 9:4993 (1970).

9. Gabbiani G., Ryan G. B., and Majno G., Presence of modified
    fibroblasts in granulation tissue and their possible role
    in wound contraction, Experientia 27:549 (1971).

10. Gabbiani G., Chaponnier C., and Huttner I., Cytoplasmic fila-
    ments and gap junctions in epithelial cells and myofibro-
    blasts during wound healing, J. Cell Biol. 76:561 (1978).

11. Majno G., Gabbiani G., Hirschel B. J., Ryan G. B., and Statkov
    P. R., Contraction of granulation tissue in vitro: similarity
    to smooth muscle, Science 173:548 (1971).

12. Ryan G. B., Cliff W. J., Gabbiani G., Irle C., Statkov P. R.,
    and Majno G., Myofibroblasts in an avascular fibrous tissue,
    Lab. Invest. 29:197 (1973).

13. Ryan G. B., Cliff W. J., Gabbiani G., Irle C., Montandon D.,
    Statkov P. R., and Majno G., Myofibroblasts in human gran-
    ulation tissue, Hum. Pathol. 5:55 (1974).

14. Laemmli U. K., Cleavage of structural proteins during the
    assembly of the head of bacteriophage T4, Nature 227:680
    (1970).

15. Gabbiani G., The myofibroblast: A key cell for wound healing and
    fibrocontractive diseases, Progr. Clin. Biol. Res. 54:182
    (1981).

16. O'Farrel P. H., High resolution two-dimensional electrophoresis
    of proteins, J. Biol. Chem. 250:4007 (1975).

17. Gabbiani G, Schmid E., Winter S., Chaponnier C., de Chasto-
    nay C., Vandekerckhove J., Weber K., and Franke W. W.,
    Vascular smooth muscle cells differ from other smooth muscle
    cells: predominance of vimentin filaments and a specific
    -type actin, Proc. Natl. Acad. Sci., U.S.A. 78:298 (1981).

18. Madden J. W., and Peacock E. E., Studies on the biology collagen
    during wound healing. III. Dynamic metabolism of scar
    collagen and remodelling of dermal wounds, Ann. Surg. 174:
    511 (1971).

19. Bailey A. J., Bazin S., and Delaunay A., Changes in the nature
    of the collagen during development and resorption of gran-
    ulation tissue, Biochem., Biophys. Acta 328-383 (1973).

20. Hansen T. M., Collagen development in granulation tissue as
    compared with collagen of skin and aorta from injured and
    non-injured rats, Acta Path. Microbiol. Scand. A, 83:721
    (1975).

21. Bailey A. J., Sims T. J., Le Lous M., and Bazin S., Collagen
    polymorphism in experimental granulation tissue, Biochem.
    Biophys. Res. Commun. 66:1160 (1975).

22. Babbiani G., Le Lous M., Bailey A. J., Bazin S. and Delaunay A.,
    Collagen and myofibroblasts of granulation tissue. A
    chemical, ultrastructural and immunologic study, Virchows
    Arch. (Cell Pathol) 21:133 (1976).

23. Bazin S., Le Lous M., Duance V. C., Sims T. J., Bailey A. J., Gabbiani G., D'Andiran G., Pizzolato G., Browski A., Nicoletis C., and Delaunay A., Biochemistry and histology of the connective tissue of Dupuytren's disease lesions, Eur. J. Clin. Invest. 10:9 (1980).

24. Narayanan A. S., Page R. C., and Kuzan F., Collagens synthesized in vitro by diploid fibroblasts obtained from chronically inflamed connective tissue, Lab. Invest. 39:61 (1978).

25. Vande Berg J. S., Rudoph R., and Woodward M., Comparative growth dynamics and morphology between cultured myofibroblasts from granulating wounds and dermal fibroblasts, Am. J. Pathol. 114:187 (1984).

26. Somers K. D., Dawson D. M., Wright G. L., Leffell M. S., Rowe J. J., Bluemink G. G., Vande Berg. J. S., Gleischman S. H., Devine C. J., and Horton C. E., Cell culture of peyronie's disease plaque and normal penile tissue, J. Urol. 127:585 (1982).

27. Azzarone B., Failly-Crepin C., Daya-Grosjean J., Chaponnier C., and Gabbiani G., Abnormal behavior of cultured fibroblasts from nodule and non-affected aponeurosis of Dupuytren's disease, J. Cell. Physiol. 117:353 (1983).

28. Gabbiani G., Hirschel B. J., Ryan G. B., Statkov P. R., and Majno G., Granulation tissue as a contractile organ. A Study of structure and function, J. Exp. Med. 135:719 (1972).

29. Adelstein R. S., Conti M. A., Johnson G. S., and Pastan I., Isolation and characterization of myosin from cloned mouse fibroblasts, Proc. Natl. Acad. Sci. U.S.A. 69:3693 (1972).

30. Bray D., and Thomas C., The actin consert of fibroblasts, Biochem. J. 147:221 (1975).

31. Gabbiani G., Majno G., and Ryan G. B., The fibroblast as a contractile cell: the myofibroblast, in: "Biology of Fibroblast", E. Kulonen and J. Pikkarainen, eds., Academic Press, London (1973).

32. Lazarides E., and Weber K., Actin antibody: the specific visualization of actin filaments in non-muscle cells, Proc. Natl. Acad. Sci U.S.A. 71:2268 (1974).

33. Painter R. G., Sheetz M., and Singer S. J., Detection and ultrastructural localization of human smooth muscle myosin-like molecules in human non-muscle cells by specific antibodies, Proc. Natl. Acad. Sci, U.S.A. 72:1359 (1975).

34. Weber K., and Groeschel-Stewart U., Antibody to myosin: the specific visualization of myosin-containing filaments in non-muscle cells, Proc. Natl. Acad. Sci., U.S.A. 71: 4561 (1974).

35. James D. W., and Taylor J. F., The stress developed by sheets of chick fibroblasts in vitro, Exp. Cell Res. 54:107 (1969).

36. Ariyan S., Enriquez R., and Krizek T. J., Wound contraction and fibrocontractive disorders, Arch. Surg. 113:1034 (1978).

37. Baur P.S., Larson D. L., and Stacey T. R., The observation of myofibroblasts in hypertrophic scars, Surg. Gynecol. Obstet. 141:22 (1975).

38. Baur P.S., Parks D. H., and Larson D. L., The healing of burn wounds, Clin. Plast. Surg. 4:389 (1977).

39. Dabelsteen E., and Kremenak C. R., Demonstration of actin in the fibroblasts of healing palatal wounds, Plast. Reconstr. Surg. 62:429 (1978).

40. Grimaud J. A., and Borojevic R., Myofibroblasts in hepatic schistosomal fibrosis, Experientia 33:890 (1977).

41. Guber S., and Rudolph R., The myofibroblast, Surg. Gynecol. Obstet. 146:641 (1978).

42. Larson D. L., Abston S., Willis B., Linares H., Dobrkovsky M., Evans E. B., and Lewis S. R., Contracture and scar formation in the burn patient, Clin. Plast. Surg. 1:653 (1974).

43. Madden J. W., On "the contractile fibroblast", Plast. Reconstr. Surg. 52:291 (1973).

44. Peacock E. E., Wound contraction and scar contracture, Plast. Reconstr. Surg. 62:600 (1978).

45. Roland J., Fibroblaste et myofibroblaste dans le processus granulomateux, Ann. Anat. Pathol. 21:37 (1976).

46. Rudolph R., and Woodward M., Spatial orientation of microtubules in contractile fibroblasts in vivo, Anat. Rec. 191:169 (1978).

47. Rudolph R., Guber S., Suzuki M., and Woodward M., The life cycle of the myofibroblast, Surg. Gynecol. Obstet. 145:389 (1977).

48. Rudolph R., McClure W. J., and Woodward M., Contractile fibroblasts in chronic alcoholic cirrhosis, Gastroenterol. 76: 704 (1979).

49. Zimman O. A., Robles J. M., and Lee J. C., The fibrous capsule around mammary implants: an investigation, Aest. Plast. Surg. 2:217 (1978).

50. Benjamin S. P., Mercer R. D., and Hawk W. A., Myofibroblastic contraction in spontaneous regression of multiple congenital mesenchymal haematomas, Cancer 40:2343 (1977).

51. Chiu H. F., and McFarlane R. M., Pathogenesis of Dupuytren's contracture: a correlative clinical-pathological study, J. Hand Surg. 3:1 (1978).

52. Feiner H., and Kaye G. I., Ultrastructural evidence of myofibroblasts in circumscribed fibromatosis, Arch. Path. Lab. Med. 100:265 (1976).

53. Fisher E. R., Paulson J. D., and Gregorio R. M., The myofibroblastic nature of the uterine plexiform tumor, Arch. Pathol. Lab. Med. 102:477 (1978).

54. Gabbiani G., and Majno G., Dupuytren's contracture: fibroblast contraction? An ultrastructural study, Am. J. Pathol. 66:131 (1972).

55. Gokel J. M., and Hubner G., Occurrence of myofibroblasts in the different phases of morbus Dupuytren (Dupuytren's contracture), Beitr. Pathol. 161:166 (1977).

56. Hueston J. T., Hurley V. J. and Whittingham S., The contracting fibroblast as a clue to Dupuytren's contracture, The Hand 8:10 (1976).

57. Katenkamp D., and Stiller D., Cellular composition of the so-called dermatofibroma (histiocytoma cutis), Virchows Arch. (Pathol. Anat.) 367:325 (1975).

58. Madden J. W., Carolson E. C., and Hines J., Presence of modified fibroblasts in ischemic contracture of the intrinsic musculature of the hand, Surg. Gynecol. Obstet. 140:509 (1975).

59. Schwarzlmuller B., and Hofstadter F., Fibromatose der Schilddrusenregion. Eine elektronenmikroskopische und enzymhistochemische studie, Virchows Arch. (Pathol. Anat.) 377: 145 (1978).

60. Weathers D. R., and Campbell W. G., Ultrastructure of the giant-cell fibroma of the oral mucosa, Oral Surg. 38:550 (1974).

61. Wirman J. A., Nodular fasciitis, a lesion of myofibroblasts. An ultrastructural study, Cancer 38:2378 (1976).

62. Bhathal P. S., Presence of modified fibroblasts in cirrhotic livers in man, Pathology 4:139 (1972).

63. Rudolph R., McClure W. J., and Woodward M., Contractile fibroblasts in chronic alcoholic cirrhosis, Gastroenterol. 76:704 (1979).

64. Nagle R. B., Evans L. W., and Reynolds D. G., Contractility of renal cortex following complete ureteral obstruction, Proc. Soc. Exp. Biol. Med. 148:611 (1975).

65. Churg A. M., and Kahn L. B., Myofibroblasts and related cells in malignant fibrous and fibrohistiocytic tumors, Hum. Pathol. 8:205 (1977).

66. Stiller D., and Katenkamp D., Cellular features in desmoid fibromatosis and well-differentiated fibrosarcomas. An electron microscopic study, Virchows Arch. (Pathol. Anat.) 369:155 (1975).

67. Vasudev K. S., Harris M., A sarcoma of myofibroblasts. An ultrastructural study, Arch. Pathol. Lab. Med. 102:185 (1978).

68. D'Andiran G., and Gabbiani G., A metastasizing sarcoma of the pleura composed of myofibroblasts, in: "Progress in Surgical Pathology", Vol. II, C. M. Fenoglio and M. Wolff, eds., Masson Publishing U.S.A. Inc., New York (1980).

69. Pollard T. D., and Wihing R. R., Actin and myosin and cell movement, CRC Crit. Rev. Biochem. 2:1 (1974).

SECTION 2

CELL SURFACE INTERACTIONS AND BIOTOLERANCE: BASIC ASPECTS

ENDOTHELIUM AS A HAEMOCOMPATIBLE SURFACE

J.D. Pearson

Section of Vascular Biology
Clinical Research Centre
Harrow, Middx. HA1 3UJ, U.K.

INTRODUCTION

Vascular endothelial cells form a monolayer that, except at a few specialised sites, continuously lines blood vessels and constitutes the primary interface between the bloodstream and all extravascular tissue.

Classically regarded as a selective but passive barrier allowing the transfer of essential nutrients from the bloodstream, endothelium is now known to be an important modulator of many aspects of vascular homeostasis, including blood coagulation, blood platelet aggregation, leukocyte emigration and vascular tone. Our increasing knowledge of how endothelial cell functions are involved in these processes derives in large part from the relatively recent ability to isolate homogeneous populations of vascular endothelial cells and grow them in culture.[1]

This in vitro approach has been restricted by two factors, although current research is directed towards overcoming these limitations. Firstly, until a few years ago no-one had succeeded in culturing microvascular endothelium and even now it is not a routine procedure.[2] Endothelial cells in microvessels expose a much greater luminal surface area to the blood, which also has a much lower flow rate than in large vessels; functional interactions between endothelium and blood components are thus far more probable in microvascular beds. Large vessel endothelial cells, which in vivo may be functionally differentiated from those in the microvasculature, are therefore not necessarily the most appropriate cells with which to study such interactions in vitro. Secondly, endothelial cells in culture may not express the same differentiated phenotype as they did

53

Table 1.   Endothelium, coagulation and fibrinolysis

Procoagulant activities

> Synthesis of basement membrane components
> Synthesis of Factor V
> Binding sites for Factors Va, IXa, Xa
> Expression of Tissue Factor (thromboplastin)

Anticoagulant activities

> Activation of Protein C
> Inactivation of circulating thrombin
> Synthesis and secretion of plasminogen activator

in vivo.[3] Nonetheless, substantial advances in our understanding of
endothelial cell biology have been made from experiments on
endothelial cells in vitro, in conjunction with studies of perfused
vascular beds.

In this short review, I have attempted to summarise the
contributions that endothelial cells make to the processes of blood
clotting and blood platelet aggregation. More general recent reviews
of endothelial cell function can be found in refs. 4-6.

ENDOTHELIUM AND BLOOD COAGULATION

The major interactions between endothelial cells and the blood
clotting and fibrinolytic cascades are listed in Table 1. Under
normal conditions endothelial cells present a non-coagulant surface
to flowing blood. When a blood vessel is damaged, clotting can occur
by either the extrinsic or the intrinsic pathways. The intrinsic
pathway is initiated by the activation of circulating Factor XII
(Hageman Factor). Intact endothelium does not interact with Factor
XII (ref. 7), but subendothelial structures - notably collagen,
which is one of the components synthesised by endothelial cells as a
constituent of their basement membrane[8] - activate Factor XII when
blood is exposed to them.[9] The intrinsic arm of the coagulation
cascade leads to the activation of Factor IX to IXa, which (as shown
in the simplified scheme in Figure 1) binds Factor VIIIa in the
presence of phospholipid to form a complex that activates Factor X.
Factor Xa, in association with Factor Va and phospholipid (PL), then
converts prothrombin to thrombin, which in turn cleaves fibrinopeptides from fibrinogen, allowing polymerisation of the resulting
fibrin to form a clot. Once the intrinsic pathway has been triggered,
endothelial cells contribute in several ways to the cascade of
reactions. They bind Factors IXa, Xa, possibly VIIIa, and synthesize
Factor V and present Va at their surface.[10-13] Endothelial cells
also express suitable phospholipid moieties to form the ternary

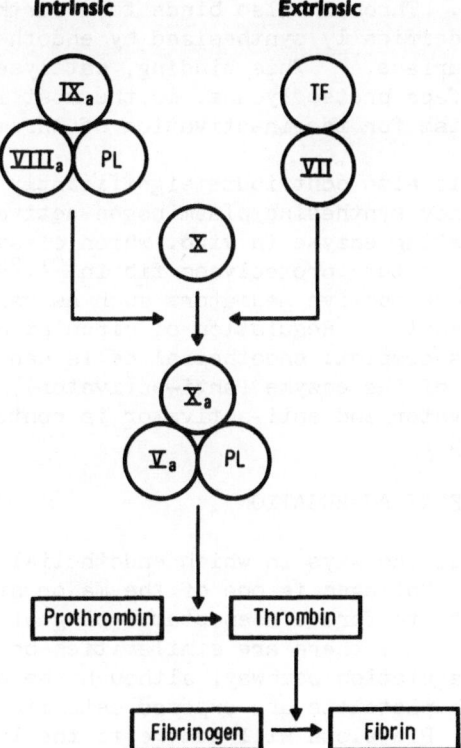

Fig. 1.   Schematic outline of final reactions of the
           coagulation cascade.

complexes that activate Factor X and prothrombin,[14] so that the
luminal surface of endothelium may be a favoured site for these
reactions.

        Endothelial cells, unlike several non-vascular cell types, do
not normally express Tissue Factor (TF, thromboplastin) and therefore
do not initiate the extrinsic arm of the coagulation system.[15]
Active thrombin, however, binds to endothelial cells and stimulates
the expression of this lipoprotein complex, which forms a receptor
for Factor VII and thus amplifies the activation of Factor X (ref.16).

        There are at least two further interactions between thrombin and
endothelium which affect the blood coagulation system; in these cases
the consequences are anticoagulant. Thrombin binds to a specific
endothelial cell surface component, named thrombomodulin, that
enhances by several orders of magnitude the ability of thrombin to
proteolytically cleave another circulating component, Protein C
(ref. 17). Thus activated, Protein Ca functions to inactivate

Factors Va and VIIIa. Thrombin also binds to a further protein, antithrombin III, specifically synthesised by endothelial cells and expressed at their surface.[18] This binding, catalysed by heparin-like endothelial surface proteoglycans, is the most important physiological mechanism for the inactivation of thrombin.[19,20]

Endothelial cells also contribute significantly to fibrinolysis of a formed clot. They synthesize plasminogen activator and are the source of the circulating enzyme in vivo, which cleaves plasminogen to plasmin, and this in turn proteolyses fibrin.[21,22] Stimulators of secretion include vasoactive mediators such as vasopressin, bradykinin or noradrenalin. Regulation of circulating plasminogen activator activity is complex: endothelial cells can concomitantly secrete an inhibitor of the enzyme (anti-activator), but how the balance between activator and anti-activator is controlled is as yet poorly understood.[21]

ENDOTHELIUM AND PLATELET AGGREGATION

Tabe 2 summarises the ways in which endothelial cells regulate platelet activation. Collagen is one of the major stimuli inducing platelets to aggregate to form a haemostatic plug after vessel wall damage has occurred. Thus there are similarities between this process and the intrinsic coagulation pathway, although the physical form of the collagen to which platelets are exposed determines the degree of platelet activation. Platelets will adhere to the less fibrillar collagens present in basement membrane; this adhesion is catalysed by Factor VIIIvw, a protein specifically synthesised by endothelial cells and which can be released in response to stimuli – including thrombin.[23-25] Platelet adhesion to interstitial fibrillar collagens, however, additionally induces the secretion of platelet products (notably ADP and thromboxane A_2) that cause platelet-platelet aggregation and further secretion.[26] Activation of the coagulation cascade also directly stimulates platelet aggregation, since thrombin is a potent platelet stimulant.

Endothelial cell products can therefore be involved in promoting the thrombotic process to prevent extravascular leakage of blood constituents, but endothelial cells also possess a variety of properties involved in the inhibition of thrombus formation, which may be regarded as regulators limiting the undesirable process of intravascular clotting.

Shortly after the discovery that stimulated platelets mobilise intracellular arachidonic acid from phospholipids and convert it primarily to thromboxane A_2 – a potent, labile, aggregating agent[27] – it was demonstrated that stimulation of eicosanoid metabolism in endothelial cells led to the synthesis and secretion of prostacyclin (prostaglandin I_2) – a potent, labile, anti-aggregating agent.[28-31] Although prostacyclin was for a time considered as a circulating

Table 2. Endothelium and platelet function

Prothrombotic activities

    Synthesis of basement membrane components
    Synthesis of Factor VIIIvw

Antithrombotic properties

    Synthesis of prostacyclin
    Non-thrombogenic luminal surface
    Inactivation of circulating thrombin
    Inactivation of adenine nucleotides
    Inactivation of serotonin, prostaglandins

hormone, constantly released by endothelial cells to prevent intra-vascular platelet adhesion or aggregation, it is now recognised that circulating levels of prostacyclin are too low to affect platelet function directly.[32] Prostacyclin is much less potent as an inhibitor of platelet adhesion than of aggregation,[33] and the low affinity of platelets for the surface of endothelial cells compared to certain other cell types is not related to endothelial prostacyclin production,[34] suggesting that specific elements of the membrane composition of the endothelial cell are responsible for its non-thrombogenic luminal surface. The release of prostacyclin from endothelial cells is triggered by a variety of vasoactive mediators,[6,35] including platelet stimulants such as thrombin and ADP,[30,36] so that anti-aggregating levels of prostacyclin are produced locally to inhibit thrombus growth.

We have studied prostacyclin release from endothelium in response to several agonists, using endothelial cells cultured on microcarrier beads and packed into small columns, which allowed us to follow the time course of release in detail.[36-37] We have also used this technique to characterise the P_2 purinoceptor recognising ADP at the endothelial cell surface. This receptor also responds to ATP, and can be clearly distinguished from endothelial ectoenzymes hydrolysing ATP or ADP (see below).[36] Whilst investigating the effects of proteinases on endothelial cells, several of which induce prostacyclin release, we found that similar concentrations of proteinases also induce the selective release of adenine nucleotides from endothelial cells, in concentrations sufficient to affect platelet aggregation.[38,39] Thus the damaged vessel wall can be a source of circulating ADP, contributing to the initiation of platelet adhesion and aggregation.

Endothelial cells regulate the circulating levels of several vasoactive and platelet-active mediators, by three main mechanisms:

(1) inhibitors present at the cell surface; (2) ectoenzymes at the cell surface; (3) uptake and intracellular metabolism. Each of these processes contributes to the control of platelet aggregation. Thrombin is inactivated by binding to endothelial antithrombin III, as described above. Endothelial ectoenzymes include peptidases such as angiotensin converting enzyme, and nucleotidases.[40] It has been known for many years that adenine nucleotides are efficiently catabolised on passage through capillary beds, and we have characterised in detail the three ectoenzymes involved, which sequentially break down ATP→ADP→AMP→adenosine.[41,42] Pro-aggregatory ADP is thus converted to anti-aggregatory adenosine. This compound is inactivated by the third process noted above; endothelial cells efficiently transport adenosine from the circulation and convert it to intracellular ATP.[43,44]

Endothelial cell metabolism of extracellular nucleotides and uptake of adenosine apparently occurs similarly in capillary beds and in large vessel endothelial cells.[45] Localisation of the inactivation in vivo to the microvasculature therefore seems to be a consequence of flow and surface area considerations. In contrast, while endothelial cells almost completely inactivate pro-aggregatory prostaglandins (e.g. PGE_2) or serotonin, by similar processes of specific carrier-mediated uptake and intracellular metabolism, on single passage through capillary beds (notably the lung), endothelial cells in large vessels apparently lack the membrane carriers to remove these compounds.[46] This suggests that in these cases microvascular endothelium behaves in vivo qualitatively differently from large vessel endothelium, as postulated in the Introduction.

CONCLUSIONS

Endothelial cells express specific products, secrete specific mediators, and possess specific enzymic mechanisms that are involved in many of the reactions of blood coagulation and platelet aggregation. The endothelium normally presents to the bloodstream a haemocompatible interface that does not induce coagulation or platelet aggregation. When either of these processes is initiated, whether by vessel wall damage or by intravascular mediator generation, the appropriate endothelial cell reactions are activated to modulate the vascular response. The properties of the endothelium are thus integral to the initial generation of a fibrin clot and to its subsequent dissolution, and to the initial formation of a haemostatic plug and limitation of its intravascular growth.

REFERENCES

1.   E.A. Jaffe, R.L. Nachman, C.G. Becker and C.R. Minick, Culture of human endothelial cells derived from umbilical veins. Identification by morphologic and immunologic criteria. J. Clin. Invest., 52: 2745 (1973).

2. B.R. Zetter, Culture of capillary endothelial cells, in:
    "Biology of Endothelial Cells," E.A. Jaffe, ed., Martinus
    Nijhoff, Boston, p.14 (1984).
3. J.L. Gordon and J.D. Pearson, Responses of endothelial cells to
    injury, in: "Pathobiology of the Endothelial Cell," H.L. Nossel
    and H.J. Vogel, eds., Academic Press, New York, p.433 (1982).
4. H.L. Nossel and H.J. Vogel, eds., "Pathobiology of the
    Endothelial Cell", Academic Press, New York (1982).
5. E.A. Jaffe, ed., "Biology of Endothelial Cells," Martinus
    Nijhoff, Boston (1984).
6. J.L. Gordon and J.D. Pearson, Biology of the vascular
    endothelium, in: "Haemostasis and Thrombosis," A.L. Bloom and
    D.P. Thomas, eds., Churchill Livingstone, Edinburgh, 2nd edtn.
    (1985).
7. G.M. Rodgers, C.S. Greenburg and D.A. Shuman, Characterization
    of the effects of cultured vascular cells on the activation of
    blood coagulation, Blood, 61:1155 (1983).
8. H. Sage, P. Pritzl and P. Bornstein, Secretory phenotypes of
    endothelial cells in culture: comparison of aortic, venous,
    capillary and corneal endothelium, Arteriosclerosis, 1:427
    (1981).
9. G.D. Wilner, H.L. Nossel and E.C. LeRoy, Activation of Hageman
    factor by collagen, J.Clin.Invest., 47:2608 (1968).
10. R.L. Heimark and S.M. Schwartz, Binding of coagulation factors
    IX and X to the endothelial cell surface, Biochem. Biophys.Res.
    Commun., 111:723 (1983).
11. D.M. Stern, D. Drillings, H.L. Nossel, A. Hurlet-Jensen,
    K.S. La Gamma and J. Owen, Binding of factors IX and IXa to
    cultured vascular endothelial cells, Proc.Natl.Acad.Sci. (USA),
    80:4119 (1983).
12. I. Maruyama, H.H. Salem and P.W. Majerus, Coagulation factor Va
    binds to human umbilical vein endothelial cells and accelerates
    protein C activation, J.Clin.Invest., 74:224 (1984).
13. T.J. Cerveny, D.N. Fass and K.G.Mann, Synthesis of coagulation
    factor V by cultured aortic endothelium, Blood, 63:1467 (1984).
14. G.M. Rodgers and M.A. Shuman, Prothrombin is activated on
    vascular endothelium by factor Xa and calcium, Proc.Natl.Acad.
    Sci.(USA), 80:7001 (1983).
15. G.M. Rodgers, G.J.J. Broze and M.A. Shuman, The number of
    receptors for factor VII correlates with the ability of cultured
    cells to initiate coagulation, Blood, 63:434 (1984).
16. T. Lyberg, K.S. Galdal, S.A. Evensen and H. Prydz, Cellular
    cooperation in endothelial cell thromboplastin synthesis,
    Br.J.Haematol., 53:85 (1983).
17. N.B. Esmon, W.G. Owen and C.T. Esmon, Isolation of a membrane-
    bound cofactor for thrombin-catalyzed activation of protein C,
    J.Biol.Chem. 257:859 (1982)

18. T.K. Chan and V. Chan, Antithrombin III, the major modulator
     of intravascular coagulation, is synthesized by human endothelial
     cells, Thrombos.Haemostas., 46:504 (1981).

19. P. Lollar and W.G. Owen, Clearance of thrombin from circulation
     in rabbits by high-affinity binding sites on endothelium. J.Clin.
     Invest., 66:1222 (1980).

20. C. Busch and W.G. Owen, Identification in vitro of an endothelial
     cell surface cofactor for antithrombin III, J.Clin.Invest., 69:
     725 (1982).

21. D.J. Loskutoff and E. Levin, Properties of plasminogen
     activators produced by endothelial cells, in: "Biology of
     Endothelial Cells," E.A. Jaffe, ed., Martinus Nijhoff, Boston,
     p.200 (1984).

22. D. Rijken, G. Wijngaards and J. Welbergen, Relationship between
     tissue plasminogen activator and the activators in blood and
     vessel wall, Thrombos.Res., 18:815 (1980).

23. K.S. Sakariassen, P.A. Bolhuis and J.J. Sixma, Adhesion of
     human blood platelets to human artery subendothelium is mediated
     by factor VIII-von Willebrand factor bound to subendothelium,
     Nature, 279:636 (1979).

24. E.A. Jaffe, L.E. Hoyer and R.L. Nachman, Synthesis of von
     Willebrand factor by cultured human endothelial cells, Proc.
     Natl.Acad.Sci.(USA), 71:1906 (1974).

25. J.H. Reinders, P.G. De Groot, M.D. Gonsalves, J. Zandbergen,
     C. Loesberg and J.A. Van Mourik, Isolation of a storage and
     secretory organelle containing von Willebrand protein from
     cultured human endothelial cells, Biochem.Biophys.Acta., 804:
     361 (1984).

26. S.A. Santoro and L.W. Cunningham, The interaction of platelets
     with collagen, in: "Platelets in Biology and Pathology 2,"
     J.L. Gordon, ed., Elsevier-North Holland, Amsterdam, p.249
     (1981).

27. M. Hamberg, J. Svensson and B. Samuelsson, Thromboxanes; a new
     group of biologically active compounds derived from prostaglandin
     endoperoxides, Proc.Natl.Acad.Sci.(USA), 72:2994 (1975).

28. S. Bunting, R. Gryglewski, S. Moncada and J.R. Vane, Arterial
     walls generate from prostaglandin endoperoxides a substance
     (prostaglandin X) which relaxes strips of mesenteric and coeliac
     arteries and inhibits platelet aggregation, Prostaglandins,
     12:897 (1976).

29. D.E. MacIntyre, J.D. Pearson and J.L. Gordon, Localisation and
     stimulation of prostacyclin production in vascular cells,
     Nature, 271:549 (1978).

30. B.B. Weksler, C.W. Ley and E.A. Jaffe, Stimulation of endothelial
     cell prostacyclin production by thrombin, trypsin and the
     ionophore A23187, J.Clin.Invest., 62:923 (1978).

31. S. Moncada, A.G. Herman, E.A. Higgs and J.R. Vane, Differential
     formation of prostacyclin by layers of the arterial wall,
     Thrombos.Res., 11:323 (1977).

32. T.A. Blair, S.E. Barrow and K.A. Waddell, Prostacyclin is not a circulating hormone in man, Prostaglandins, 23:597 (1982).

33. E.A. Higgs, S. Moncada, J.R. Vane, J.P. Caen, H. Michel and G. Tobelem, Effects of prostacyclin on platelet adhesion to rabbit arterial subendothelium, Prostaglandins, 16:17 (1978).

34. K.D. Curwen, M.A. Gimbrone Jr. and R.J. Handin, The role of prostacyclin in platelet adhesion to normal and virally transformed endothelial cells, Lab.Invest., 42:366 (1980).

35. R.I. Levin, B.B. Weksler, A.J. Marcus and E.A. Jaffe, Prostacyclin production by endothelial cells, in: "Biology of Endothelial Cells," E.A.Jaffe ed., Martinus Nijhoff, Boston, p.228 (1984).

36. J.D. Pearson, L.L. Slakey and J.L. Gordon, Stimulation of prostaglandin production through purinoceptors on cultured porcine endothelial cells, Biochem.J., 214:273 (1983).

37. J.D. Pearson, J.S. Carleton and A. Hutchings, Prostacyclin release stimulated by thrombin or bradykinin in porcine endothelial cells cultured from aorta or umbilical vein, Thrombos.Res., 29:115 (1983).

38. J.D. Pearson and J.L. Gordon, Vascular endothelial and smooth muscle cells in culture selectively release adenine nucleotides, Nature, 281:384 (1979).

39. E.C. LeRoy, A. Ager and J.L. Gordon, Effects of neutrophil elastase and other proteases on procine aortic endothelial prostaglandin I_2 production, adenine nucleotide release, and responses to vasoactive agents, J.Cell.Invest., 74:1003 (1984).

40. J.W. Ryan and U.S. Ryan, Endothelial surface enzymes and the dynamic processing of plasma substrates, Int. Rev.Exp.Pathol., 26:1 (1984).

41. N.J. Cusack, J.D. Pearson and J.L. Gordon, Stereoselectivity of ectonucleotidases on vascular endothelial cells, Biochem. J., 214:317 (1983).

42. J.D. Pearson, Ectonucleotidases of vascular endothelial cells: characterisation and possible physiological roles, in: "Cellular Biology of Ectoenzymes," G.W. Kreutzberg, M. Reddington and H. Zimmerman, eds., Springer, Berlin (1985).

43. J.D. Pearson, J.S. Carleton, A. Hutchings and J.L. Gordon, Uptake and metabolism of adenosine by pig aortic endothelial and smooth muscle cells in culture, Biochem. J., 170:265 (1978).

44. J.D. Pearson and J.L. Gordon, Nucleotide metabolism by endothelium, Ann.Rev.Physiol., 47:617 (1985).

45. P.G. Hellewell and J.D.Pearson, Metabolism of circulating adenosine by the porcine isolated perfused lung. Circulation Res., 53:1 (1983).

46. J.D. Pearson and J.L. Gordon, Metabolism of serotonin and adenosine, in: "Biology of Endothelial Cells," E.A. Jaffe ed., Martinus Nijhoff, Boston, p.330 (1984).

INTERACTION OF BLOOD, BLOOD COMPONENTS AND SURFACES

J.C. Barbenel,[1] K.J. Mynett,[1] J.M. Courtney,[1]
C.D. Forbes,[2] and G.D.O. Lowe [2]

1. Bioengineering Unit, University of Strathclyde, Glasgow
2. Dept. of Medicine, Glasgow Royal Infirmary, Glasgow

INTRODUCTION

It has been realised for many years that coagulation may be
initiated by contact between the blood and foreign surfaces, and
that such artificial surfaces and the normal vascular endothelium
have markedly different effects [1]. Interest in the interaction
between blood and foreign surfaces has received great impetus from
the development of devices which remain in contact with the blood
either continuously, e.g. cardiovascular prostheses, or intermitt-
ently, as in extracorporeal blood purification systems. The materials of all these appliances deleteriously interact with blood and
there is, as yet, no completely satisfactory biomaterial [2].

The increased investigation of blood-material interactions has
led to a greater awareness of their complexity [3]. The complex
process of blood coagulation includes contributions from platelet
activation, intrinsic, extrinsic and common pathways and the systems
involved in thrombosis inhibition and fibrinolysis [4]. The events
following blood-material contact can be considered in the same light.
There is, however, an important difference between the vascular
endothelium and foreign materials. The endothelium interface can
play an active role in thromboresistance, which is not available
in artificial surfaces. Thus, the use of devices which contain
foreign materials often require the use of anticoagulants, inhibitors of platelet aggregation or plasminogen activators.

INTERACTION OF BLOOD AND ARTIFICIAL SURFACES

The interaction of blood on artificial surfaces follows a well
established sequential series of events which involve components

of the plasma, platelets and red and white cells.

Protein Adsorption

Proteins are rapidly adsorbed onto artificial surfaces from whole blood [5], protein solutions [6,7], or plasma [8], although this may be proceeded by the adsorbtion of water and inorganic ions. The proteins in the adsorbed layer may undergo denaturation or changes in conformation [9].

The extent of protein adsorption depends upon the nature of the artificial material. Adsorption and retention is generally greater on surfaces which are hydrophobic [10]. Hydrophobic behaviour also appears to influence conformational changes after adsorption [11].

The adsorbed protein layer has a major influence on the subsequent blood-material interactions and this has led to suggestions that it may be considered as a 'conditioning' layer [12]. The adsorption of plasma proteins from blood is a selective process. Fibrinogen is adsorbed preferentially in comparison to albumin, globulin, lipoproteins and coagulation factors [8]. Alterations to the interactions between antibodies and adsorbed fibrinogen suggests that the latter may undergo a short term change in conformation [13].

Adsorbtion is also an important factor in the activation of the intrinsic coagulation pathways. Activation is initiated by the adsorption of factor XII and high molecular weight kininogen [14]. It is probable that activation of the intrinsic mechanisms leads to thrombin formation and the deposition of a monolayer of fibrin on the artificial surface [15,16], which can influence subsequent interactions.

Platelets

The reaction of platelets to artificial surfaces is of major importance in the initiation of thrombus formation, and platelet adhesion and aggregation are an invariable result of exposure of blood to foreign surfaces [17,18]. The platelet adhesion is related to protein adsorbtion [19]. Adhesion is enhanced by the presence of an adsorbed layer of fibrinogen or γ - globulin and inhibited by the presence of adsorbed albumin. It is believed that the thrombin formation that results from activation of the intrinsic mechanisms and the production of a fibrin monolayer on the surface enhances platelet adhesion [16].

Adherent platelets undergo a typical shape change, and it is extremely probable that the platelet release reaction occurs. This leads to further aggregation on the surface [20], the formation of an irregular platelet layer and the accumulation of red and white blood cells enmeshed by a fibrin network [21]. The release of several platelet constituents has been reported in response to biomaterials in clinical devices. Elevated β – thrombo-globulin (BTG) and platelet Factor 4 (PF4), levels have been reported during dialysis and after heart valve implantation [22,23].

Red and White Blood Cells

The presence of red blood cells in protein solution influences the protein adsorption process, reducing the amounts of protein adsorbed [24]. The erythrocytes also adhere to the adsorbed protein layer [19], and the haemolosis of these erythrocytes may result in the release of ADP and induce a platelet release reaction. Nevertheless Hoffman has suggested that red blood cells may play only a minor role in the process of thrombogenisis [14].

Leucocytes have procoagulant [25,26] and proaggregatory activity [27]. Once again, protein adsorption is important in the interaction between leucocytes and surfaces. Leucocyte adhesion is promoted by adsorbed γ – globulin [28], but adsorbed albumin appears to produce reduced adhesion [29,30].

Leucopenia regularly occurs in the early phases of haemodialysis when using regenerated cellulose membranes [31,32]. There is now considerable evidence that this associated with complement activation by the alternative metabolic pathway and pulmonary sequestration of leucocytes.

Mechanical Factors

The interaction of blood and foreign surfaces is influenced by the shear stress (or shear rate) to which the blood is subjected. The adhesion of platelets is increased under static no-flow conditions but, in the presence of high shear stresses, increased platelet aggregation occurs [33]. Erythrocytes are more strongly affected by shear stresses than are platelets [34], and may undergo irreversible damage if subjected to relatively low shear stresses for five minutes [35].

Leucocytes are also sensitive to shear stresses and may undergo aggregation, increased adhesion and loss of lysosonal enzymes [36,37].

The steady state composition of adsorbed albumin and fibrinogen films appear to be independent of shear rate [7].

Surface finish also influences blood-material interactions. In general, rougher surfaces are more thrombogenic than smooth ones of the same material [38].

ASSESSMENT OF BLOOD COMPATIBILITY

There is no single test which can predict the clinical performance and long term compatibility of biomaterials. In vitro assessments are, however, still of utility in developing and evaluating
candidate materials. In view of the complexity of the blood-material
interactions many possible indicators of the response of blood to
exposure to foreign materials have been suggested [39]. There is
little doubt that multiparameter assessments are necessary [40].

The response of platelets has been widely used to assess blood-
material interactions. This may be done by examining surfaces exposed to platelet rich plasma by scanning electron microscopy.
Alternatively, the change in the number of platelets remaining in
the blood exposed to the materials may be quantified. Platelet
adhesion and aggregation can be measured, using techniques developed
from the method of Wu and Hoak [41,42]. The basis of the method is
the fixation of the platelet aggregates, reduced by exposure to
foreign surfaces by the addition of a buffered EDTA and formalin
solution and the determination of the total number of platelets by
the addition of buffered EDTA alone. Comparisons with counts
obtained from unexposed blood allows the estimation of platelet
adhesion and aggregation.

The composition of the buffered solutions is critical if reliable
and reproducible platelet aggregation results are to be obtained [43,44].
Adhesion and aggregation results are generally consistent - those
materials producing least adhesion also producing least aggregation.
The functional state of the platelets exposed to the test surfaces
may also be assessed by inducing aggregation with ADP or collagen.
The results suggest that surfaces showing least aggregation and
adhesion also produce diminished functional impairment in the platelets.

The platelet release reaction plays an essential part in aggregation. The release reaction can be followed by the estimation of
βTG levels, and provides a sensitive and useful test of blood compatibility. The release of βTG can also be induced by thrombin, and
some estimate of thrombin levels must be obtained if the βTG levels
are to be correctly interpreted. Parallel estimates of fibrinopeptide A provides a suitable estimate of thrombin formation [45].

Evaluation of compatibility can be made using either platelet
rich plasma or whole blood. Recent developments make it possible
to count platelets in the presence of red and white blood cells [46].
Comparative tests, using human blood, show that the presence of
erythrocytes increases the aggregation of platelets exposed to the
foreign surfaces [47], and that this response is haematocrit dependent [48].

There has been considerable recent interest in complement activation [49]. Assessment of complement activation has been greatly enhanced by the availability of radioimmunoassays for specific complement components. Compatibility evaluation has generally been based on the measurement of C3a [50], or C5a [51].

It is important in selecting a test method to estimate compatibility, that the blood-material contact time and exposure method are standardised. The effect of anticoagulants, especially heparin, should also be considered. It is also important that species differences are recognised when interpreting and comparing test results or extrapolating the results of animal tests to clinical situations.

REFERENCES

1. J. Lister, On the coagulation of the blood, Proc. Roy. Soc., 12: 580-611, (1963).
2. K. Klinkman, The Role of Biomaterials in the Application of Artificial Organs, in: "Biomaterials in Artificial Organs" (J.P. Paul, J.D.S. Gaylor, J.M. Courtney and T. Gilchrist eds), Macmillan, London, 1-8, (1984).
3. S.D. Bruck, Properties of Biomaterials in the Physiological Environment, CRC Press, Boca Raton, (1980).
4. O.D. Ratnoff and C.D. Forbes, Disorders of Haemostasis, Grune and Stratton, Orlando, (1984).
5. R.M. Gendreau, S. Winters, R.I. Leininger, D. Fink, C.R. Hassler, R.J. Jakobsen, Fourier transform infrared spectroscopy of protein adsorption from whole blood: ex vivo dog studies, Applied Spectroscopy, 35: 353-357, (1981).
6. J.L. Brash and D.J. Lyman, Adsorption of plasma proteins in solution to uncharged hydrophobic polymer surfaces, J. Biomed. Mater.Res., 3: 175-189, (1969).
7. B.M.C. Chan and J.L. Brash, Adsorption of fibrinogen on glass: reversibility aspects, J. Colloid. Interface. Sci., 82: 217-225, (1981).
8. L. Vromen, A.L. Adams, M. Klings, G.C. Fisher, P.C. Munoz and R.P. Solensky, Reactions of formed elements of blood with plasma proteins at interfaces, Ann. New York Acad. Sci., 283: 65-76, (1972).
9. W.H. Lee and P. Hairston, Structural effects of blood proteins at the gas-blood interface, Fed. Proc. 30: 1615-1620, (1971).
10. B.D. Ratner, Biomedical applications of hydrogels, in: Biocompatibility of Clinical Implant Materials, Vol:2, D.F. Williams ed. CRC Press, Boca Raton, 145-175, (1981).
11. D.R. Absolon, A.W. Neumann, W. Zingg and C.J. van Oss, Thermodynamic studies of cellular adhesion. Trans ASAIO, 15: 152-156, (1979).
12. J.L. Brash, Protein adsorption and blood interactions in: Biocompatible polymers, metals and composites, M. Szycher ed. Technomic, Lancaster (Pa), 35-52, (1983).

13. L. Vromen, A.L. Adams, G.C. Fisher and P.C. Munoz, Interaction of high molecular weight kininogen, Factor XII and fibrinogen in plasma at interfaces, Blood, 55: 156-159, (1980).

14. A.S. Hoffman, Blood-biomaterials interactions, in: Biomaterials, Interfacial Phenomena and Applications, S.L. Cooper and N.A. Peppas, eds. Am. Chem. Soc. Washington, D.C., 3-8, (1982).

15. D.F. Waugh and D.J. Baughman, Thrombin adsorption and possible relations to thrombus formation, J. Biomed. Mater.Res., 3: 145-164, (1969).

16. H.Y.K. Chuang, P.E. Crowther, S.F. Mohammed and R.G. Mason, Interactions of thrombin and antithrombin III with artificial surfaces, Thrombosis Research, 14: 273-282, (1979).

17. R.G. Mason, The interactions of blood haemostatic elements with artificial surfaces, Prog. in Haemostasis and Thrombosis, 1: 141-164, (1972).

18. R.G. Mason, S.F. Mohammed, H.Y.K. Chuang and P.D. Richardson, The adhesion of platelets to subendothelium, collagen and artificial surfaces, Seminars in Thrombosis and Haemostasis, 3: 98-116, (1976).

19. J. Feijen, Thrombogenesis caused by blood - foreign surface interactions, in: Artificial Organs, R.M. Kenedi, J.M. Courtney, J.D.S. Gaylor and T. Gilchrist eds., Macmillan, London, 235-247, (1977).

20. H.R. Baumgartner, R. Muggli, T.B. Tschopp and V.T. Turitto, Platelet adhesion, release and aggregation in flowing blood, Thromb. Haemostas., 35: 124-138, (1976).

21. E.W. Salzman, J. Landon and D. Brier, Surface-induced platelet adhesion, aggregation and release, in: The Behaviour of Blood and its Components at Interfaces, L. Vroman and E.F. Leonard eds. New York Acad. Sci., New York, 114-127, (1979).

22. A.J. Adler and G.M. Berlyne, Thromboglobulin and platelet factor 4 levels during haemodialysis with polyacrylonitrile, Trans. ASAIO, 4: 100-102, (1981).

23. R. Dudczkak, H. Niessner, E. Thaler, K. Lechner, K. Kletter, H. Frishauf, E. Domanig and H. Aisher, - Thromboglobulin, platelet factor 4 and fibrinopeptide A in patients with porcine and prosthetic heart valves, Proc. VIII International Congress on Thrombosis, Haemostasis, London, Thromb. Haemostas. 42: 72, (1979).

24. S. Uniyal, J.L. Brash and I.A. Degterev, Influence of red blood cells and their components on protein adsorption, Am. Chem. Soc. Advances in Chemistry, 199: 277-292, (1982).

25. J. Niemetz, Coagulant activity in leucocytes, J. Clin. Invest., 51: 307-313, (1972).

26. H.I. Saba, J.C. Herion, R.I. Walker and H.R. Roberts, The procoagulant activity of granulocytes, Proc. Soc. for Experimental Biol. and Med., 142: 614-620, (1973).

27. M.J. Harrison, P.R. Emmons and J.R. Mitchell, The effect of white cells on platelet aggregation, Thrombosis et Diathesis Haemorrhagica, 16: 105-121, (1966).

28. A.L. Adams, G.C. Fischer and L. Vromen, The complexity of blood at simple interfaces, J. Colloid. Interface. Sci., 65: 468-478, (1978).

29. E.W. Salzman, E.W. Merill, A. Binder, C.R.W. Wolf, T.P. Ashford and W.G. Austin, Protein platelet interactions on heparinised surfaces, J. Biomed. Mater. Res., 3: 69-81, (1969).

30. D.J. Lyman, K.G. Klein, J.L. Brash and B.K. Fitzinger, The interaction of platelets with polymer surfaces, Thrombosis et Diathesis Haemorrhagica, 23: 120-128, (1970).

31. P.R. Craddock, J. Fehr, A.P. Dalmasso, K.L. Bringham and H.S. Jacob, Haemodialysis leucopenia: pulmonary vascular leuco-stasis resulting from complement activation by dialyser cellophane membranes, J. Clin. Invest, 59: 879-888, (1977).

32. P.C. Farrell, Biocompatibility aspects of extracorporeal circulation, in: Biomaterials in Artificial Organs, J.P.Paul, J.D.S. Gaylor, J.M. Courtney and T. Gilchrist eds., Mac-millan, London, 342-350, (1984).

33. S. Yu, J.G. Latour, B. Marchandise and M. Bois, Shear stress - induced changes in platelet reactivity, Thromb. Haemostas. 40: 551-560, (1978).

34. P.D. Richardson, S.F. Mohammed and R.G. Mason, Flow chamber studies of platelet adhesion at controlled, spatially varied shear rates, Proc. Eur. Soc. Art. Org., 4: 175-188, (1977).

35. B.N. Nanjappa, H.K. Chang and C.A. Glomski, Trauma to the erythrocyte membrane associated with low shear stress, Biophys. J., 13: 1212, (1973).

36. T.S. Dewitz, T.C. Hung, R.R. Martin and L.V. McIntire, Mechan-ical trauma in leukocytes, J. Lab. Clin. Med., 90: 728-736, (1977).

37. D.J. Stockwell and L.V. McIntyre, Alterations of polymorpho-nuclear neutrophil leukocyte response to shear stress ex-posure in vitro, in: Biomaterials: Interfacial phenomena and applications. S.L. Cooper and N.A. Peppas eds., Am. Chem. Soc., Washington D.C., 209-219, (1982).

38. J.F. Hecker and L.A. Scandrett, Roughness and thrombogenicity of the outer surfaces of intravascular catheters, J. Biomed. Mater. Res., 19: 381-395, (1985).

39. C.D. Forbes, Thrombosis and artificial surfaces, Clinics in Haematology, 10: 653-668, (1981).

40. J.D. Andrade, D.L. Coleman, P. Didisheim, S.R. Hanson, R. Mason and E. Merrill, Blood-materials interactions - 20 years of frustration, Trans, ASAIO, 27: 659-662, (1981).

41. K.K. Wu and J.C. Hoak, A new method for quantitative detection of platelet aggregates in patients with arterial insufficiency, Lancet, 2: 924-926, (1974).

42. K.K. Wu and J.C. Hoak, Increased platelet aggregates in patients with transient ischaemic attacks, Stroke, 6: 521-524, (1975).

43. S.K. Bowrt, J.M. Courtney, C.R.M. Prentice and J.T. Douglas, Utilisation of the platelet release reaction in the blood com-patibility assessment of polymers, Biomaterials,5: 289-292,(1984).

44. S.K. Bowry, C.R.M. Prentice and J.M. Courtney, A modification
    of the Wu and Hoak method for the determination of platelet
    aggregation and platelet adhesion, Thrombo. Haemostas., 53:
    381-385, (1985).
45. S.K. Bowry, J.M. Courtney, C.R.M. Prentice, M. Travers, G.S.
    Brown, G.D.O. Lowe and C.D. Forbes, Biomaterial Assessment:
    measurement of platelet adhesion and aggregation in: Biomat-
    erials in Artificial Organs. J.P. Paul, J.D.S. Gaylor, J.M.
    Courtney and T. Gilchrist eds: Macmillan, London, 219-227,
    (1984).
46. A.R. Saniabadi, G.D.O. Lowe, C.D. Forbes, C.R.M. Prentice and
    J.C. Barbenel, Platelet aggregation studies in whole human
    blood, Thromb. Res., 30: 625-632, (1983).
47. A.R. Saniabadi, G.D.O. Lowe, J.C. Barbenel and C.D. Forbes,
    A comparison of spontaneous platelet aggregation in whole
    blood, with platelet rich plasma: additional evidence for the
    role of ADP, Thromb. Haemostas., 1: 115-118, (1984).
48. A.R. Saniabadi, G.D.O. Lowe, J.C. Barbenel and C.D. Forbes,
    Further studies on the role of red blood cells in spontaneous
    platelet aggregation, Thromb. Res., 38: 225-232, (1985).
49. G.A. Herzlinger, Activation of complement by polymers in contact
    with blood, in: Biocompatible polymers, metals and composites,
    M. Szycher ed: Technomic, Lancaster (Pa), 89-101, (1983).
50. D.E. Chenoweth, Biocompatibility of haemodialysis membranes:
    evaluation with C3a anaphylatoxin radioimmunoassays. Am. Soc.
    Art. In. Org., 7: 44-49, (1984).
51. J. Breillatt and W.J. Dorson, Haemocompatibility studies. Am.
    Soc. Art. Int. Org., 7: 57-63, (1984).

CELLULAR ASPECTS OF BIOTOLERANCE

Trevor Rae

Orthopaedic Research Unit
University of Cambridge Clinical School
Addenbrookes Hospital
Cambridge, CB2 2QQ

INTRODUCTION

This contribution concentrates on just one aspect of the theme
of the conference and that is the interaction of cells with foreign
surfaces. There are many instances in which cells come into contact
with foreign materials and include the following examples. The
inhalation of particulates and their subsequent interaction with
cells of the lung. This may be a result of industrial exposure to
asbestos and silica or the domestic exposure to glass fibres used in
loft insulation for example. In contrast to this some foreign materials are purposely implanted in the body to augment or replace the
function of natural structures. Examples include various metals,
polymers and ceramics used in orthopaedic, cardio-vascular, dental
and plastic surgery.

Other routes of contact between cells and foreign materials
include those which take place outside the body. Interactions may
occur between blood cells and materials used in the construction of
equipment for extracorporial circulation used during dialysis and bypass surgery. Blood cells from donors may also contact foreign
polymer surfaces during prolonged storage. More recently, the
storage of sperm and ova provides another example in which cells
contact foreign materials.

Probably the most studied interaction of cells with foreign
surfaces is that of cells in monolayer culture. Most of such observations, however, would almost certainly have been done unwittingly
during the course of some other study.

One extremely important practical application of studies on the interaction between cells and foreign materials is in the screening of materials for possible use in the body. For the purpose of this presentation most examples will be drawn from studies on materials for use in orthopaedic applications. Before considering in more detail the cellular aspects of such screening a few comments should be made on the current terminology used in this field. Most frequently, reference is made to studies on the biocompatibility of biomaterials for use in the body. More accurately, this should be referred to as biotolerance of xenomaterials for use in the body. The use of the latter phrase is more appropriate for the following reasons. First, the majority of materials currently in use (and probably those still to be proposed) offer only some degree of toler- ance not compatibility, that is there is a "grudging coexistence" between cells and materials rather than an "integrated harmonious coexistence" which is conveyed by the use of the word compatibility. Second, almost without exception, materials currently in clinical use are of a non-biological origin and are more accurately referred to as xenomaterials. However, it is probably now too late to change the usage of these words as biocompatibility and biomaterials have become an entrenched part of the terminology of this particular field.

Materials which are implanted in the body can be divided into two broad categories, that is those which are "inactive" or "passive" and which are not intended to evoke any specific tissue reactions within the body to fulfil their function. The temptation is to describe such materials as inert; they are not, cells invariably recognise them and react in a distinctive manner.

In contrast to the above, another group of materials which have been developed depend upon recognition by the host tissues in order to fulfil their function. Porous orthopaedic implant materials for instance require the specific response of bone ingrowth in order to obtain a degree of attachment to the skeleton. Thus the cellular response which is required for the success of these two categories of implant materials are different and screening procedures should reflect these differences in end use.

In the context of screening materials for adverse reactions it is meaningless therefore to specify which test methods should be employed without reference to the circumstances under which the material is to be employed clinically. The problem of screening is compounded further by the absence of a definitive set of tests, biotolerance has to be established in a stepwise manner by carefully choosing methods to establish that undesirable cell reactions are not induced. On what criteria should these tests be based? This can be answered by reference to the cellular response to foreign materials, some of these interactions can form the basis for such test methods.

A number of diverse responses can occur when cells interact with

a foreign surface, some are summarised in Figure 1. Some are direct
and immediately obvious such as the cytotoxic effects of a component
of a polymer or alloy. Some of the effects are direct but only
become apparent after long periods of time, examples include the
induction of tumours and fibrotic lesions of the lung after exposure
to asbestos. Other effects may be indirect in which, for instance,
the result of the interaction compromises part of the host defense
mechanisms and only becomes manifest when such defenses are challenged by an infection. The ramifications of the interaction of cells
with foreign surfaces with specific reference to the cellular
response to orthopaedic materials is considered below.

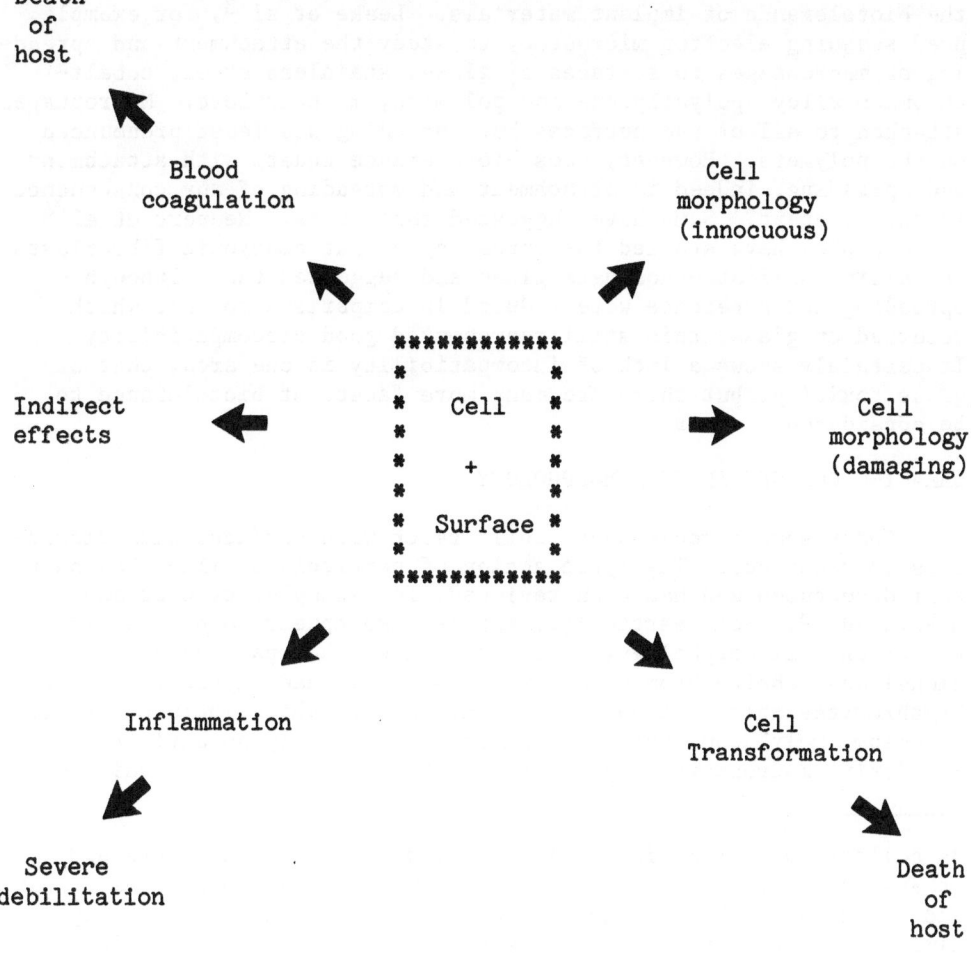

Fig. 1. Possible ramifications of contact between cells and foreign
materials

INNOCUOUS CHANGES IN CELL MORPHOLOGY

Probably one of the most extensively studied interactions between cells and a foreign surface is that which occurs when cells are grown in culture. It is probably fair to say that most of these studies will have been done so unwittingly, the majority being done for some other reason. Some of the earliest direct observations, of cells in culture were carried out in the 1920's by Strangeways [1] and Lewis [2]. Canti and Strangeways were probably the first to produce a time lapse film* of the migration of cells from an explant and their subsequent behaviour in culture. Although most cells exhibit a considerable degree of tolerance to their artificial substrata in that they carry out most of their natural processes their morphology is atypical. The direct observation of the interaction of cells with surfaces has been utilised by some people to assess the biotolerance of implant materials. Leake et al [3], for example, used scanning electron microscopy to study the attachment and spreading of macrophages to surfaces of glass, stainless steel, cobalt-chromium alloy, polyethylene and polymethylmethacrylate. Macrophages attached to all of the surfaces but spreading was least pronounced on the polymers. However, does biotolerance equate with attachment and spreading, indeed is attachment and spreading of any consequence in this context? Some have suggested that it is. Neupert et al [4] for example, have studied the spreading of rat embryonic fibroblasts on calcium-silicate-phosphate glass and suggested that although spreading and adherence were reduced in comparison to that which occurred on glass, this still represented good biocompatibility. It certainly shows a lack of incompatibility in one area, that of gross toxicity, but there are many more facets of biotolerance to be considered however.

DAMAGING CHANGES IN CELL MORPHOLOGY

Under some circumstances cells react with surfaces with disastrous consequences. The lytic action of particulate silica has been well documented and has been reviewed, for example, by Uber and McReynolds [5]. Some particulate metals also appear to have a direct effect on cell morphology. Rae [6] has shown that particulate cobalt, nickel and cobalt-chromium alloy are strongly haemolytic for human erythrocytes whereas a number of other metals which include titanium, chromium, molybdenum and tantalum showed virtually no activity. Similarly, macrophages exposed to these particulates were quickly

---

*A collection of time lapse films made by Canti, Strangeways and others during the period 1928-35 is held by the National Film Archive and is known as the Strangeways Collection. A grant from Dunlop Limited to make a video recording of one of these films (Cells in Tissue Culture, 1927) for presentation at the conference is gratefully acknowledged.

killed by cobalt, nickel and the alloy but remained unaffected by
other metals [7]. Although the damaging metals are relatively soluble
in biological fluids, which accounts for their longer-term toxicity
to macrophages and fibroblasts, soluble metal did not cause haemolysis.
It appears, therefore, that some materials in particulate form have
the potential to damage cells directly.

Foreign material in particulate form may also possess other
peculiar properties not shown by the bulk parent material and which
are relevant in the present context. First, owing to the large
ratio of surface area to mass, the solubility of possibly toxic
components will be exaggerated. Second, the increased surface area
also leads to an increased possibility of direct interaction between
the cell surface and foreign surface. Third, particulate material
of less than about 5μm can be phagocytosed, once inside the cell
any direct damaging actions or release of toxic agents are likely
to be more pronounced.

CELL TRANSFORMATIONS

One possible consequence of the interaction of cells with a
foreign surface is the formation of a tumour. Numerous factors
may be involved in this process of so-called solid state carcinogenesis and include the chemical composition and physical state of
the implanted material. Some of the earliest observations were
made by Turner [8]. Phenolformaldehyde (Bakelite) discs which were
implanted subcutaneously in rats produced a number of fibrosarcomas.
Similarly, Oppenheimer et al [9] showed that cellophane discs induced
fibrosarcomas and rhabolomyosarcomas. The role of the physical form
of the implant in the induction of tumours has been reported by
Nothdurft [10]. Three different materials showed an almost identical
potency for the induction of tumours when in the form of unperfor-
ated discs. All materials progressively lost their carcinogenic
property when the physical form of the material which was implanted
was either a perforated disc or a powder. The results are summarised
in Figure 2. Several authors have reported on the carcinogenic
potential the various physical forms of polyethylene, the details

	Polystyrene	Hydrocellulose	Polyvinylchloride
Unperforated discs	79% (37/47)	82% (57/72)	79% (75/95)
Perforated discs	41% (40/98)	72% (44/61)	43% (33/76)
Powder	0	0	0

Figure 2   Influence of Physical Form on Tumour Production

are summarised in Figure 3. The carcinogenic potential of implant materials has been extensively reviewed by Pedley et al [11], and the more general phenomenon of solid state carcinogenic by Bischoff and Bryson [12] and by Ott [13].

Although tumours have occurred at the site of implants in humans there is little evidence to suggest that there was a direct link between the two. This has been recently reviewed by Hamblen and Carter [14].

INFLAMMATION

Inflammation is often one of the most visible signs of response to injury or of bio-intolerance. Examples of severe chronic inflammation have occurred around grossly corroded bone plates for instance. Even around implants which have not visibly corroded, however, varying degrees of inflammation are present.

A common feature of the tissue response is the absence of the "immune element" of inflammation. That is, polymorphonuclear leuco-cytes, macrophages, foreign body giant cells and fibrobalsts are seen at various stages but only rarely are lymphocytes and plasma cells seen. A possible explanation for the presence of the latter cells in any quantity is the manifestation of hypersensitivity to components of the implanted materials or the presence of a bacterial infection.

Form	Tumour formation	Authors
20mm x 0.015mm discs	40% (35/87)	Bates & Klein [15] (1966)
Solid	35% (7/20)	Carter & Roe [16] (1969)
Shredded	25% (5/20)	
Powder	3% (1/30)	Oppenheimer et al [17] (1961)

Fig. 3. Influence of Physical Form on Tumour Induction by Polyethylene

Materials vary in their capacity to induce inflammation, two
clinically-used materials which provide examples of two degrees of
activity are carbon fibre and high density polyethylene (HDP).

    Carbon fibre when injected into the knee joints of mice caused
an unremarkable response over a period of two weeks to one year.
Rushton and Rae [18] described that the reactive process consisted
of a very localised slight synovial hyperplasia with only a minimal
infiltration of mononuclear cells. This contrasted to the reactions
seen to HDP in which there was a rapid (within 2 weeks) response of
large numbers of macrophages which surrounded the particulate HDP.
The densely-packed cells were difficult to distinguish from multi-
nucleated giant cells which were a common feature at later time
intervals. At longer intervals the 'islands' of granulation tissue
were surrounded by fibroblasts and a thin fibrous capsule.

    In an effort to discover some of the biochemical events
associated with the above cellular manifestations of inflammation the
pattern of release of marker enzymes has been investigated. Broadly
speaking, the effect that particulate material which is phagocytosed
may have on macrophages is summarised in Figure 4. Lactate dehydrogenase (LDH) an enzyme marker of cell membrane integrity found in the

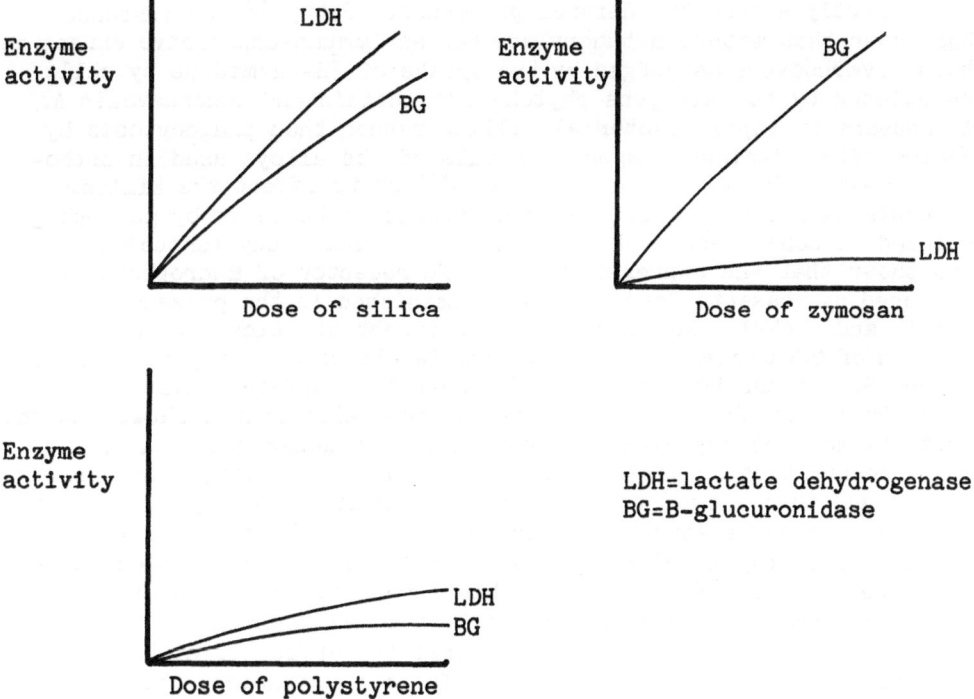

Fig. 4. The pattern of enzyme release of macrophages exposed to
        three materials with different inflammatory potential.

soluble cell fraction and B-glucuronidase (B-G) a lysosomal enzyme marker are both released into the supernant by materials which are extremely damaging such as silica. Known inflammatory agents such as zymosan and some forms of asbestos cause a specific release of lyso-somal markers but not of LDH. Other innocuous particles such as polystyrene cause no release of either group of markers. Rae (unpublished) has used the approach to study the inflammatory potent-ial of titanium and its aluminium-vanadium alloy.

INDIRECT EFFECTS

In some cases the cellular aspects of biotolerance only become manifest when the compromised response is tested by a third party. For example, the infection of orthopaedic prostheses can arise many years after replacement of the joint. It may be facilitated by a localised defect in the foreign defense mechanisms of the host brought about as a direct result of the presence of the implanted foreign material. Such a defect will thus pass unnoticed until challanged by an infection.

The factors which pre-dispose to late infection are largely unknown. The accumulation of corrosion products from alloys and components of polymers, especially unpolymerised methyl methacrylate may directly affect the defence processes. Petty [19] for instance has shown that methyl methacrylate has an immunosuppressive effect on human lymphocytes as judged by the uptake of ^{3}H-thymidine by cells stimulated by the mitogens phytohaemagglutinin and concanavalin A. It appears to impair bacterial killing rather than phagocytosis by neutrophils. Various component metals of the alloys used in ortho-paedic alloys have been shown by Rae [20] not to affect the killing of bacteria but to reduce the phagocytosis of bacteria by macrophages exposed to cobalt and nickel. A more detailed study (unpublished) has shown that the expression of the Fc receptor of macrophages as measured by rossette formation was diminished in the presence of cobalt and nickel. An example of the effect of nickel at a concen-tration of 0.5 umole/ml with various levels of antibody is shown in Figure 5. It can be seen that for control, untreated cells, the distribution of the number of attached red cells starts skewed to the left (ie most macrophages have only a few attached red cells). As the concentration of antibody increased the distribution becomes normal and then skewed to the right (i.e. most macrophages have more than ten red cells attached to them). At lower levels of antibody the corresponding metal treated samples retain a distribution which is skewed to the left. However, this effect diminishes as the anti-body concentration rises and the highest level there is an appreciable degree of protection against the effect of nickel. A similar picture is seen for cobalt. Table 1 summarises the results for various concentrations of metal and antibody, the protective effect of serum is clearly visible.

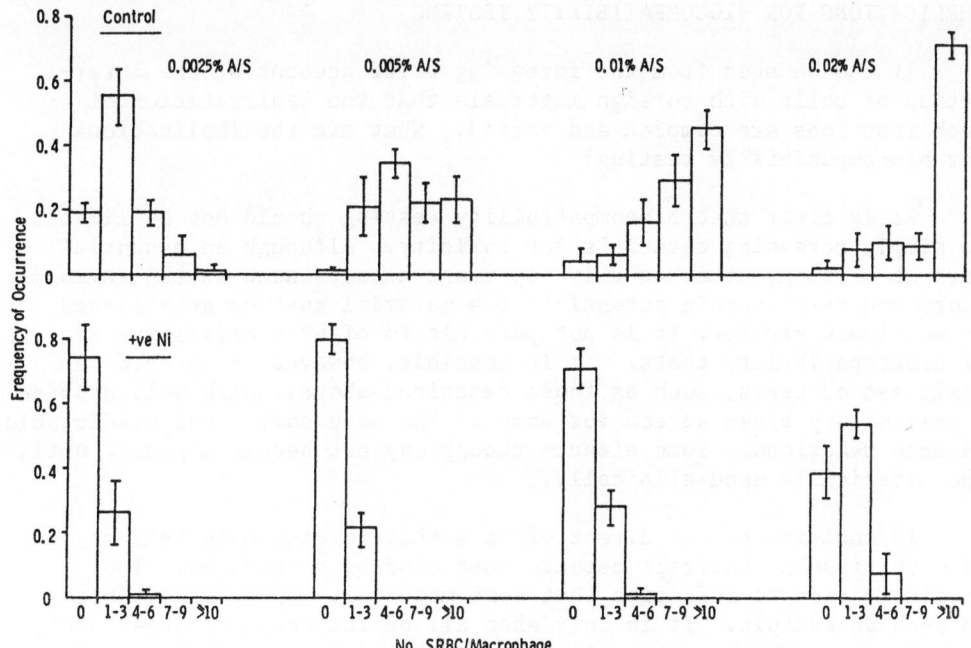

Fig. 5. Frequency distribution of sheep red blood cells attached to macrophages. Effect of 0.5 umole/ml nickel chloride for 18 hr.

Table 1 Summary of the percentage of macrophages forming rosettes in the presence of either nickel or cobalt at various concentrations of metal and anti-serum. (More than 6 attached red cells was scored as a rosette)

Metal conc.	Antiserum concentration		
umole/ml	0.005	0.01	0.02
Ni 0.25	99.1	83.4	93.4
0.50	78.7	84.4	95.8
1.00	44.7	81.5	101.6
Co 0.25	79.5	92.8	100.1
0.50	57.5	86.4	94.1
1.00	45.1	76.1	85.1

IMPLICATIONS FOR BIOCOMPATIBILITY TESTING

It can be seen from the foregoing brief account of the interaction of cells with foreign materials that the manifestation of
such reactions are complex and varied. What are the implications
for biocompatibility testing?

It is clear that biocompatibility testing should not be limited
to simply screening materials for toxicity. Although an essential
part of testing, other equally important aspects such as the inflammatory and carcinogenic potential of a material must be established.
As mentioned earlier, it is not possible to offer a definitive set
of biocompatibility tests. It is possible, however, to provide a
basic set of tests, such as those described above, which will provide
a preliminary broad screen for some of the more common and predictable
adverse reactions. Some effects though may not become apparent until
the material is used clinically.

In addition to the direct effects that foreign materials may
have on tissues, indirect aspects must also be considered. The
predisposition to infection that some materials appear to exhibit
is such an example. It is only when all of the cellular aspects of
biotolerance have been established that it will be possible to offer
a complete set of tests for the routine screening of materials.

REFERENCES

1. T.S.P. Strangeways, Observations on the changes seen in living
        cells during growth and division, Proc. Roy. Soc. B., 94:
        137, (1922).
2. W.H. Lewis, Observations on cells in tissue culture with dark
        field illuminations, Anatom. Rec., 26: 15, (1923).
3. E.S. Leake, M.J. Wright and A.G. Gristina, Comparative study of
        the adherence of alveolar and peritoneal macrophages, and of
        blood monocytes to methyl methacrylate, polyethylene, stain-
        less steel and Vitallium, J. Reticuloendothel. Soc., 30 : 403,
        (1981).
4. G. Neupert, V. Thieme, H. Hofmann and G. Berger, Adhesion, spread-
        ing and growth of animal cells on the surface of glass ceramic
        Ap40 - a contribution to the cell compatibility of dental
        permanent hard tissue implants, Exp. Path. 25 : 51, (1984).
5. L.U. Uber and R.A. McReynolds, Immunotoxicology of silica, CRC
        Critical Reviews in Toxicology, 10 : 303, (1982).
6. T. Rae, The haemolytic action of particulate metals, J. Path.
        125 : 81, (1977).
7. T. Rae, A study on the effects of particulate metals of ortho-
        paedic interest on murine macrophages in vitro, J. Bone
        Joint Surgery, 57B: 444, (1975).
8. F.C. Turner, Sarcoma at sites of subcutaneously implanted
        Bakelite discs in rats, J. Natl. Canc. Inst. 2 : 81, (1941).

9. B.S. Oppenheimer, B.T. Oppenheimer and A.P. Stout, Sarcoma
    induced in rats by implanting cellophane, Proc. Soc. Expt.
    Biol. Med., 67 * 33, (1948).
10. H. Nothdurft, Experimentelle sarkomauslosung durch eingeheilte,
    Fremdkorper, Strahlentherpie, 100 : 192, (1956).
11. R.B. Pedley, G. Meachim and D.F. Williams, Tumor induction by
    implant materials, in: Fundamental Aspects of Biocompatibility
    (Volume II), D.F.Williams etc., CRC Press Inc. Boca Raton,
    Florida.
12. F. Bischoff and G. Bryson, Carcinogenesis through solid state
    surfaces, Progr. Exp. Tumor Res. 5 : 85, (1964).
13. G. Ott, Fremdkorpersarkome, Springer-Verlag, Berlin, (1970).
14. D.L. Hamblen and R.L. Carter, Sarcoma and joint replacement,
    J. Bone Joint Surgery, 66B: 625, (1984).
15. B.R. Bates and M. Klein, Importance of a smooth surface in
    carcinogenesis by plastic film, J. Natl. Canc. Inst. 37 : 145,
    (1966).
16. R.L. Carter and F.J.C. Roe, Induction of sarcomas in rats by
    solid and fragmented polyethylene: experimental observations
    and clinical implications, Br. J. Cancer, 23 : 401, (1969).
17. E.T. Oppenheimer, M. Willhite, I. Danishefsky and A.P. Stout,
    Observations on the effects of powdered polymer in the
    carcinogenic process, Canc. Res., 21 : 132, (1961).
18. N. Rushton and T. Rae, The intra-articular response to particulate
    carbon fibre reinforced high density polyethylene and its
    constituents: an experimental study in mice, Biomaterials,
    5 : 352, (1984).
19. W. Petty, The effect of polymethylmethacrylate on bacterial
    phagocytosis and killing by human polymorphonuclear leukocytes,
    J. Bone Joint Surgery, 60A : 752, (1978).
20. T. Rae, The action of cobalt, nickel and chromium on phagocytosis
    and bacterial killing by human polymorphonuclear leucocytes;
    its relevance to infection after total joint arthroplasty,
    Biomaterials, 4 : 175, (1983).

SURFACE PROPERTY REQUIREMENTS FOR BIOMATERIALS

G.W. Hastings

Bio-Medical Engineering Unit, (N. Staffs Polytechnic and
Staffordshire Health Authority), Medical Institute
Hartshill, Stoke-on-Trent, England

Implantation

The quest for the ideal biomaterial necessitates consideration
of the type of reaction caused by insertion into living tissues.
The early definitions of ideality covered short-term (toxicological)
and medium/long-term (immunological, carcinogenic) effects but did
not adequately define the specific materials properties required.

This resulted to some extent from what may be termed an engineering materials approach. A material was selected on the basis
of engineering properties and then the implantation of the foreign
material was used to adjudge that material acceptable for use.
When something went wrong a change was made. Thus, based on the
"non-stick" properties of polytetrafluroethylene (ptfe) Charnley [1]
devised a new operation for endoprosthetic replacement of the hip.
The abraded particles caused severe tissue reaction including abdominal granulomas, ("teflonomas") arising from transport and accumulation of debris at sites other than the implant and this use was
abandoned. Yet ptfe is acceptable in other applications not involving production of wear particles.

At this stage there was limited information about in vivo behaviour of implant materials and the setting up of adequate predictive studies was not possible. The surgery of hip replacement,
however, gave hope of satisfaction and so continued with a different
material, ultra high molecular weight polyethylene (UHMWPE), again
on the basis of engineering properties.

However, the initial use of joint endoprostheses was in a
significantly old aged group of patients and the major experience

83

is still from those of 60 years and over. With the increased pressure to reduce the age for operation in order to retain as much residual mobility as possible there may be factors other than engineering adequacy and local tissue response to be considered. Already
the in vivo degradation of polyethylene has been reported on the
basis of radioactivity in the urine of rats in which radio-labelled
polymer had been implanted [2]. The effect of saline aging on polymers following gamma irradiation was shown to be an increase in the
rate of property degradation compared with samples stored in a dess-
icator [3]. Therefore, radiation sterilisation may damage the polymer
before it is implanted and then implantation will increase the degradation effect. If the dose of radiation received is greater than
the usual 2.5 Mrad or the dose rate is too high the degradation will
be increased. Leachable components increase with degradation. Reports are now appearing on the degradation of UHMWPE used in joint
endoprostheses [4,5]. Black [6] has referenced some of the attempts
to improve the properties of the material highlighting the incomplete
satisfaction with performance.

As the duration of implantation and the in vivo demands on the
material increase so also may the incidence of observable adverse
sequelae.

Examples may be drawn from other areas to illustrate the way
in which the engineering materials approach has dominated implant
materials application. Vascular prosthesis development was regarded
in terms of textile technology, the requirement being for non-kinkable
tubes having limited porosity. Fibre morphology seemed to be the
major factor until the attempts to produce non-thrombogenic surfaces
led to more detailed knowledge of surface chemistry. The polymer
chemists have consequently had an increasing influence as the emphasis has changed from bulk material to surface chemistry.

In metals too, it is very rare now to see a grossly corroded
retrieved implant. Metals have been improved as much due to the
effects of the Standards Organisations as to fundamental research
programmes [7]. Studies on metal implantation have been designed to
improve performance by eliminating corrosion or fatigue failure or
have studied the tissue phenomena immediately adjacent to the implanted material. For frankly non-toxic materials the value of
this approach may be questioned since the observations may be quite
unrelated to patient use and its long-term effectiveness taking all
factors into consideration. However, the wider consequences are
being realised and there is a move away from phenomenology to a
study of mechanisms and the chemistry associated with the processes [8]. This is highly important when implant design is being
changed for mechanistic reasons, e.g. cementless fixation with
porous or textured surfaces. The continuing interfacial reactions
in these cases may prove to be more important than the immediate
adjacent tissue changes, in the same way that the political and

economic changes following a battle may in the long-term be more
important and enduring than the immediate battle-field carnege.

Biocompatible?

The complex of inter-relationships which follows the insertion
of inanimate material into living tissue requires a definition of
acceptable parameters by which suitability or otherwise can be
determined. In general a mechanistic approach has been used in
which the gross effect produced is noted during a period of animal
implantation often of short duration. Following this the material
may be said to be biocompatible but the deficiency in this is that
little is said concerning the physical and chemical events which
are the actual determinants of its behaviour and the test may have
little relationship to the intended use. A parallel has been drawn
between this approach and attempting to understand the events occurring during the battle of Waterloo by visiting the scene of battle
some days later [9]. It is not possible to understand the process of
decision making, the motivation of the individuals concerned and
the detail of the various actions will have been lost. Such "a
posteriori" examination may contribute nothing to the prevention
of future hostilities.

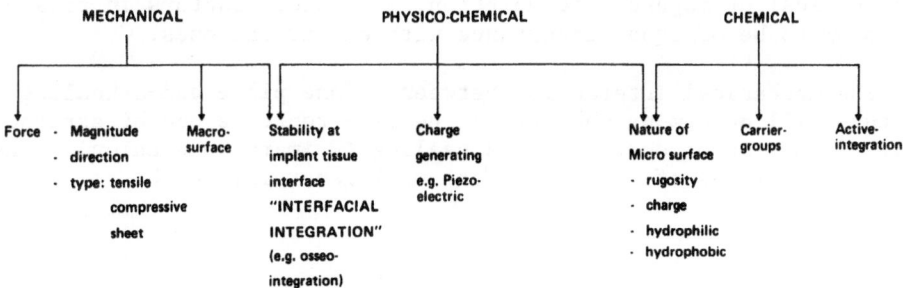

FIGURE 1

For similar reasons the study of "implant histology" is of
limited value for assurance of long-term in vivo performance or in
providing information on how that performance can be enhanced.

Black [9] believes that the term biocompatibility is inappropriate
because it implies a biased value judgement. It may indeed sometimes be based on inadequate evidence of a purely observational kind.
The declaration of biocompatibility may in fact be derived from a
concept of intertness which, in practise, will never be found. For
this reason the term "biological performance" proposed by Black may
be preferred together with the complimentary parts of material

response and host response. This approach was already adopted by
the author and colleague [10] who referred to a "benign reactivity",
recognising that a reaction took place but that it was an acceptable
level. For example, sutures elicit a response but it is accepted
in view of the overall benefit conferred by their use.

The term biological performance or tissue acceptability has a
further advantage, whilst recognising that an artificial material
does not perform biologically in the strictest sense, in that it
encourages the concept of site-specific use. This in turn directs
attention to what factors are different and hence to ways to adapt
materials or to control reactions. The reactions at the interface
will be the determining factors and these will now be considered
under different headings and related to the surface properties.

Mechanical

For many implant systems the mechanical interactions will be
the most important, although not exclusively so since all types of
interaction are interdependent. It is true for orthopaedic implants.
Bone growth itself and remodelling is known to be determined by
mechanical factors arising from the various loading systems that are
applied to living bone. This has been developed into a philosophy
for the complete treatment of bone fractures as developed by the
Swiss AO Group [11]. Indeed, in the treatment of ununited fractures
by electrical or magnetic stimulation, mechanical factors in treatment seem to be of equal importance with electrical ones.

The mechanical interaction between a bone plate and a healing
fracture will determine whether the healing route is one of external
callus formation or cortex-cortex healing (Primary bone union). The
surface properties of the bone plate will generally not be so

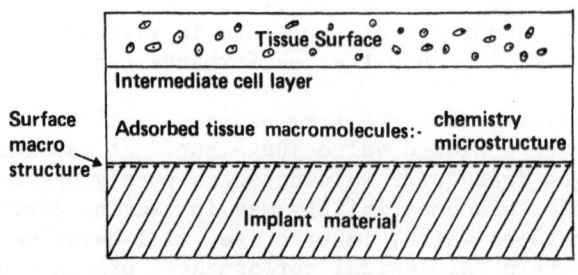

FIGURE 2

important although it is found that the carbon fibre composites used
in the series of forearm fractures by the author and colleagues [12]
appear to enhance bone growth around the plate. This is most likely
to be a response to the carbon fibre rather than to the resin since
carbon fibre is known to promote fibrous tissue responses. The next
step is to see to what extent this growth can be directed or controlled.

     For a joint prosthesis inserted into the bone the question of
interface reactions is more significant. The unresolved problem in
joint replacement surgery remains that of long-term loosening.
Although acrylic cement appears to be the means of choice for fixation of prosthesis components there is much interest in cementless
fixation [13]. The success of this lies in achieving close apposition
of new bone to the prosthesis and hence depends on surface properties.
Since bone growth will only take place at an interface where there
is no relative movement this is a problem of design. The implant
surface should ideally provide for only compressive or tensile
loading at the implant/bone interface. Much useful information has
derived from ceramic dental implants [14] where alumina ceramic
implants have been found to achieve firm fixation to bone. Various
forms of surface structure have been developed but it is generally
a macroscopic structuring which is provided. Several of the hip
prostheses used without cement follow similar principles but others
use a more finely textured surface achieved by sintering metal
spheres on to the stem. In reviewing the results from these porous
surfaced prostheses [15] the incidence of pain after about twelve
months implantation has been noted in a significant proportion of
patients. Most probably this is a loosening effect produced by the
shear forces and inadequate support at the interface. A different
result is claimed for the coating of a porous polymer composite,
"Proplast$_R$" polytetrafluorethylene (ptfe) carbon fibre material.
A ptfe-alumina composite version has also been introduced. This
type of surface has lower modulus than the metal and the load trans-
mission is said to be improved. The nature of this surface combines
porosity in the hydrophobic polymer with the more hydrophilic carbon
fibre and the resultant surface energy is such that tissue growth is
encouraged. This is an interesting concept in providing a physical
means of attachment to tissue (porosity) with a surface energy
favourable for cell attachments.

     Most attention has been given to the means for providing for
mechanical attachment by surface rugosity although biologically
active glasses (see later) applied as coating to othopaedic implants
introduce the idea of chemical activity. Since the actual nature of
the mechanical interaction is not yet fully understood developments
proceed along largely empirical lines. There is need to understand
how mechanical forces are meditated into cellular effect. If it is
by electrical means (piezo electric, streaming potentials) then it
may be possible to apply this to the design of really inter-active

mechanical interfaces which would develop renewable cell adhesion
and growth.

Physico-Chemical

Moving from purely mechanical considerations to those of physico-
chemical nature, the principle of osseo-integration has been identified by Branemark and colleagues for titanium implants inserted into
bone [16]. This is characterised by a total lack of a cellular layer
between the implant and bone, with bone protoglycans right up to the
implant surface and collagen fibres through this and onto the implant
surface. Mechanical loading, or rather its absence in the early
stages of repair is shown by the group to be of great importance.
If interface stresses are kept low, osseo-integration occurs and
subsequent loading (after 3 months) can take place. The cells in
effect "see" an oxide layer which may be likened to a ceramic or an
oxide layer modified by protein adsorption, but it is the primary
surface which initiates the sequence of events leading to bone cell
adhesion. It is, however, the secondary surface to which cells are
attached and this must be the biocompatible surface in the sense
that the macromolecules adsorbed are presented in their native form
to the cells seeking attachment. All surfaces exposed to a biological medium will undergo some modification but only in certain
circumstances will the interface conversion be favourable. Most of
the investigations have been on polymer surfaces, arising from the
interest in developing blood compatible materials.

The significance of the oxide layer on the metal surface and
its ceramic-like characteristics is yet to be understood. Alumina
ceramic also appears to be capable of direct bonding into bone and
has been used as dental implants and accetabular cups. Other
materials can also become osseo-integrated including polymethyl
methacrylate and, in the authors' experience Caron Fibre re-inforced
epoxy composite bone plates. It is certainly an important observation that close interlocking of tissue and implants can occur with
smooth surfaced implants as well as with roughened or textured
surfaces.

Whenever there is inadequate stabilisation leading to movement
or the loads at the interface are too great or improperly directed,
soft tissue interposition will occur. Stabilisation and load factors
are closely inter-related. In these circumstances osseo-integration
is not observed.

Various schemes have been proposed to account for the desired
"non-stick" properties with respect to blood cells. This is almost
the reverse of the bone requirement where strong cell attachment is
needed. In general it is the lack of control of reactions at the
interface that leads to failures. It should be remembered that the
biomechanical factors may often overshadow the physico-chemical

ones and no matter how good a surface is, movement across the interface, or excessive load patterns will predispose to failure.

However, there still remains the need to define and control the
nature of the interface and its interactions and the control of
surface properties is a significant part of this. It was said above
that the proteins or other macromolecules adsorbed should be in an
undergraded form and the implant will then produce a normal biological response from the host. Thus, the aim is to produce or develop
normal tissue at the interface of the sort that would be present if
the implants were not there. The interface should respond to
stimuli in the way that most tissue does. Whether the implant
material can be expected to behave in this way itself is at present
a counsel of perfection but should be the long-term aim. In the
absence of these ideal materials control of the interface by surface
property adjustment and, of equal importance, careful attention to
post-operative regimes that will promote development of the interface
should receive the fullest attention.

1st generation	Bulk polymer properties - engineering materials approach	Joint replacements Major implants UHMW polyethylene Acrylics
2nd generation	Surface deposition	Blood contact Heparin bonding
3rd generation	Biodegradable and hydrogels	Sutures, contact lenses, drug carriers. Polylactates, PolyHema
4th generation	Composites	Fracture fixation Carbon fibre resin Polysulphone Degradable matrix
5th generation	Specific activity	Controlled release systems. Blood contact Functional implants

FIGURE 3: THE CHANGING ROLE OF BIOMEDICAL POLYMERS.

A unifying principle must emerge if this work is to be ultimately successful. Some have sought this via the hydrophobic/hydrophilic relationships between materials. This produces anomalies in results. One of the most promising approaches has been that of Baier who classified cell growth phenomena on surfaces in terms of surface energy [17]. This certainly gives a base line for progress, but others have concentrated on equilibrium water content which seems to explain anomalous behaviour in a satisfactory way [18]. The cell type specific nature of bioactive surfaces has been noted by other workers [19], the differences being attributed to differences in cell membrane nectin concentrations.

However, to develop biomaterials that do not activate immune and other responses requires considerably greater knowledge of physico-chemical properties of biomaterials than is currently available. Attention has turned to complement activation factors and it is hoped that this will lead to more tolerable biomaterials.

Chemicals

There are purely chemical factors which also influence biomaterials behaviour. The progress in drug delivery/release systems has led to polymers which have controllable degradation rates or which give controllable diffusion profiles through the bulk material. The pattern for a targettable polymer-drug system is now well known and there have been many advances in bonding pharmacologically active agents to polymer chains [20]. The possibility for specific site reactions is now a reality particularly where monoclonal antibodies can be part of the system. This requires polymers that degrade to acceptable human metabolities.

An area largely ignored by biomaterials scientists is that of inorganic-organic reactions. Much of this work is in the area of calcified tissue research and among the pioneers Neuman and Neuman showed the close relationships between collagen and bone mineral depositions [21]. We have shown the possible importance of collagen structure in the progress of fracture healing [22] in the presence of higher than usual levels of one type of collagen in delayed unions and non-unions.

These precise relationships have received considerable study in the work of R.J.P. Williams who has studied inorganic-organic relationships in detail. He has shown that the accumulation of elements in cells is not an equilibrium process but involves complex triggers and energy barriers [23].

Surfaces of the Future

The need is to investigate whether some of the various concepts set out above can be processed into a genuine unifying principle for development of biomaterials surfaces as well as bulk properties. It is a complex issue made not the less so by the close interaction between interface reactions and biomechanics. Progress in ceramics, glasses, blood-contact materials and drug systems is giving results which should all contribute to our understanding even though at present, some of the systems requirements seem to be disparate. The sensing of external force fields, electrical, magnetic or mechanical by living organisms often involves macromolecule interactions and this too is an area of study which could be profitable for bio- materials improvement.

It is a vital question as to whether the multi-disciplinary areas can achieve the cohesion needed for this progress given the fragmented nature of much of the work. The cost of evaluation has been set out by Frisch [24] who questions whether the unfulfilled needs can be satisfied in the near future. The art of collaboration and systematic national and international programme development are likely to be as important in the near future as the science itself.

```
┌─────────────────────────────────────────────────────┐
│                                                       │
│  TISSUE e.g. BONE                                      │
│                                                       │
│  Intermediate cell layer                              │
│  Adsorbed tissue macromolecules                       │
│  Chemistry                                            │
│  Micro Structure                                      │
│                                                       │
│                                                       │
│                                                       │
│  SURFACE MACRO STRUCTURE (roughness,texture)          │
│                                                       │
│  Implant Material                                     │
│                                                       │
└─────────────────────────────────────────────────────┘
```

FIGURE 4: Onto this static interface is imposed the consequences of force application and movement. The adsorbed macromolecultes are a key element in acceptability (biological performance).

REFERENCES

1.   J. Charnley, Arthroplasty of the hip, A new operation, Lancet,
     1: 1129, (1961).
2.   B.S. Oppenheimer et al, Further studies of polymers as
     Carcinogenic agents in animals, Cancer Res., 15: 333-340,
     (1955).
3.   G.W. Hasting and A. Cooper, Irradiation of Plastics, Unpublished
     results.
4.   P. Eyerer, Degradation of UHMW polyethylene in joint endo-
     prostheses, Proc. NATO Advanced Study Institute, Marbella,
     Spain, July (1984).
5.   P. Eyerer and Y.C. Ke, Property changes of UHMW polyethylene
     hip cup endoprostheses during implantation. J. Biomed.Mater.
     Res., 18: 1137-1151, (1984).
6.   J. Black. The Failure of Polyethylene, J. Bone Jt. Surgery,
     60B: 303-306, (1978).
7.   British Standard BS3531. Surgical Implants Parts 1-16 other
     corresponding national standards and International ISO
     standards.
8.   J. Black, E.C. Martin, H. Gelman and D.M. Morris, Serum
     concentration of Chromium, cobalt and nickel after total
     hip replacement: a six month study. Biomaterials, 4: 160-
     164, (1983).
9.   J. Black, Systemic effects of biomaterials. Biomaterials,
     5: 11-18, (1984).
10. B. Bloch and G.W. Hastings, Plastics Materials in Surgery,
     2nd Edition, Charles C. Thomas, Springfield, (1972).
11. M.E. Muller, M. Allgower and H. Willenegger, Manual of Internal
     Fixation, Springer, Berlin, (1970).
12. M.S. Ali, C.H. Wynn-Jones, G.W. Hastings, T.A. French and T.Rae,
     A preliminary clinico-pathological report on the use of
     Carbon Fibre Plates in forearm fractures. Proc. Brit.Orth.
     Association, Spring Meeting (1985).
13. E. Morscher (ed), The Cementless Fixation of Hip Endoprosthesis.
     Springer, Berlin, (1984).
14. G. Heimke and W. Schulte et al, The influence of fine surface
     structure on the osseo-integration of implants. Int. J.
     Art. Organs. 5: 207-212, (1982).
15. P. Christel, Clinical Aspects of Orthopaedic Implant Surgery.
     Proc. NATO Advanced Study Institute, Applications of
     Materials Science to the Practice of Implant Orthopaedic
     Surgery. In Press (1985).
16. A. Tjellstrom, J. Lindstrom, O. Nylen, T. Albrektsson, P.I.
     Branemark, Directly bone-anchored implants for fixation of
     aural epistheses, Biomaterials, 3: 55-57, (1983).
17. R.E. Bainer and A.E. Meyer et al, Surface properties determine
     bioadhesive outcomes: Methods and results. J. Biomed. Mater.
     Res. 18: 337-351, (1984).

18. M.J. Lyndon, T.W. Minett and B.J. Tighe, Requirements for cell adhesion to synthetic polymer substrata in culture. This Symposium.

19. L. Hench, J.W. Boretos and M. Eden, Biomaterials Reliability in: Contemporary Biomaterials, Noyes Publications, Paste Ridge, N.J. U.S.A. (1984).

20. N.B. Graham and M.E. McNeill, Hydrogels for controlled drug delivery, Biomaterials, 5: 27-36, (1984).

21. W.F. Neuman and M.W. Neuman, The Chemical dynamics of bone mineral. University of Chicago Press, (1958).

22. A.M. Anderson, G.W. Hastings, T.R. Fisher, E.R.S. Ross and A. Shuttleworth, Collagen types present at human fracture sites. Injury, in press, (1985).

23. R.J.P. Williams, Physico-chemical aspects of inorganic element transfer through membranes, Phil. Trans. R. Soc., London B294: 57-74, (1981).

24. E. Frisch, The Cost Dilema-the high cost and low volume of biomaterials. In: Contemporary Biomaterials, J.W. Boretos and M. Eden (eds), Noyes Publications, 607-625, (1984).

SURFACE EMISSION AND BIOLOGICAL PROBES FOR INORGANIC INTERFACES

J.E. Davies, R.P. Hurst and N.T. Spooner

Department of Anatomy
University of Birmingham
Birmingham, B15 2TJ. England

INTRODUCTION

Inorganic materials provide solid substrata with which cells may interact in numerous natural, pathological and iatrogenic conditions. Human natural hard tissues, except dental enamel, depend upon populations of specialised cells for maintenance of their vitality and are therefore continually interactive. However, inorganic substrata may also be introduced into the body in a range of mineral dust diseases and as surgical replacements for natural tissues. The interactions of cells with materials in these last two groups may be considered to depend upon three criteria:
(i) the surface physico-chemical properties of the substratum,
(ii) the composition of the extracellular environment and
(iii) the properties of cell membranes
Each of these criteria embrace numerous aspects of the structure, composition and behaviour of the individual components which combine to create an interface in the biological environment. However, we believe that surface charge phenomena associated with both the inorganic substratum and the cell membrane will be of critical importance in interfacial behaviour.

While it has long been known that cells react to charge fields[1], little is known concerning the change in cell membrane surface-charge distribution with change in membrane morphology associated with active formation and translocation of cell membrane processes. Furthermore, while the nature of the substratum is known to alter cell morphology [2] and the latter is indicative of cell response to the former [3,4], little is understood about the mechanisms involved at such interfaces. In each of these cases the surface charge-carrier profile of an inorganic substrate may play a decisive role in the response evoked in the host tissue cells.

95

The work which we report here represents an attempt to characterize physical reactivity of inorganic solids combined with scanning
electron microscopic examination of cells in contact with such surfaces. Two in-vitro biological scenarios are dealt with; the reaction
of macrophages to mixed mineral dust particles and the colonization,
by osteoblasts, of commercially available bone substitute calcium
phosphate ceramics.

THE INORGANIC SUBSTRATUM

Ionic or insulating solids exhibit deviations from stoichiometry including point defects, dislocations, internal surfaces, voids
and inclusions. Edge dislocations are known to act as a source or
sink for vacancies and thus surface morphology and surface charge
can be intimately related. The existence of charge on the surface
and in the space-charge region (Debeye-Huckel layer) of insulators
will create compositional excursions extending over a variable depth
and modify the electrical properties of ceramic surfaces [5]. In semi-
conductors, it is known that dislocations and other bulk flaws and
impurities can provide active sites for corrosion and generate minority carriers that can induce corrosion [6]. Such charging and impurity composition of the surface will affect the behaviour of the
solid when immersed in a liquid. This type of dynamic interaction
of defects in surface behaviour is more widely reported in metals [7]
and provides an explanation, for example, of the increased corrosion
rate of materials subjected to fluctuating loads when immersed in
high dielectric environments [8]. However, unlike metals, insulators
accomodate quasi-Fermi levels, meta-stable states, within the band-
gap, which although stable at ambient temperature may be released
at activation energies well below the band-gap energy of the solid.
The relaxation of these perturbed states may play decisive role in
charge behaviour at the solid/liquid interface. The physical probe,
exoemission, we employ to characterize our inorganic substrata is
dependant upon the relaxation of such perturbed states and is briefly
described below.

EXOEMISSION

Exoemission embraces a group of charge-carrier emission phenomena which occur during the relaxation of perturbations of thermodynamic equilibria in the volume or at the surface of solids. Relaxation is accomplished by either optical or thermal stimulation
of a sample. These techniques are referred to as Optically Stimulated Exoemission (OSEE) and Thermally Stimulated Exoemission (TSEE)
respectively and have been described in detail elsewhere [9]. Emitted
electrons may be either volume or surface originating but their
release from a transition zone, between the ordered crystalline
lattice of an ionic solid and its apparent surface, will inevitably

mean that volume and surface phenomena are intimately related [10]
and probably interactive. The kinetic energies of emitted charged
particles may be far in excess of thermal activation energies due
to accelerating fields in the space-charge region [11] and other
interfacial double layers. The surface free energy therefore may be
dramatically changed by the relaxation of such perturbations. While
surface generated exoemission is associated with both desorption
phenomena and physico-chemical changes in adsorbates [12], the diffusion of ion species from the bulk due to deviations from stoichiometry, will not only contribute to the charge-carrier profile of
that solid but also be subtly changed by the pre-treatment that,
for example, a bioceramic may receive during its sintering production or sterilization procedure [13].

Exoemission differs from classical surface analysis techniques
such as X-ray Photoelectron Spectroscopy (XPS), Auger Electron
Spectroscopy (AES) or Secondary Ion Mass Spectroscopy (SIMS) which
all employ a concurrent excitation probe, since the excitation source
is removed prior to measurement. This relationship is illustrated
in Figure 1. Exoemission therefore demonstrates two unique advantages
with respect to such techniques. Firstly, since emission peaks are
associated with metastable defect structures they demonstrate a
"memory" effect in the near-surface lattice. (Relaxation of such
defects will produce a dynamic interaction at the solid/liquid interface and may be considered analogous to "conjoint-action" in metals).
Secondly, since activation energies are low, the charge-carrier information gained using these techniques will be more applicable to
an understanding of microscopic interactions at the solid/liquid
interface since changes in surface energy brought about in aqueous
environments may relax perturbations which can be monitored by exoemission measurements.

Phenomenologically, OSEE and TSEE are linked to many other
physical and physico-chemical effects. These are listed in Figure
1. While exoemission phenomena include release of various charge-
carrier species (including electrons, negative and positive ions
and excited neutrals) in the simplest case, where electrons are the
only emitted species recorded, ambient pressure gas flow electron
detectors may be employed. We have described such a system in detail [14] and it is this which has been employed to generate the emission profiles discussed below.

ELEPHANTIASIS, SOIL DUSTS AND MACROPHAGES

Elephantiasis is a disfiguring and debilitating disease caused
by long-term lymphstasis resulting in chronic oedema with skin and
subcutaneous hypertrophy. It is generally associated with filarial
infection. A non-filarial form of elephantiasis has been reported
in Africa, Central and South America. Here, people contract elephantiasis in the absence of Wucheraria bancrofti and Brugia malayi

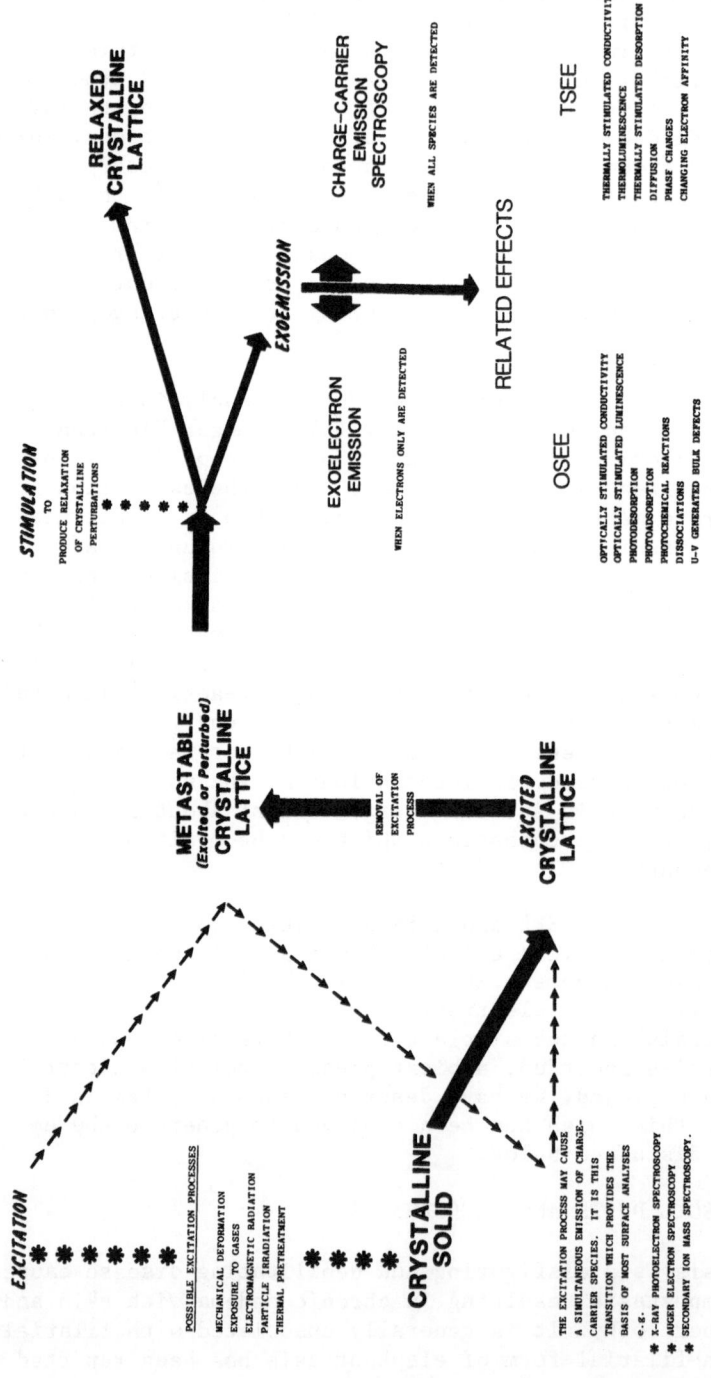

Fig. 1. Relationships of Exoemission to other physical and chemical phenomena. During excitation of a crystalline solid species may be emitted. The detection of these forms the basis of most surface analyses. After removal of the excitation, relaxation of lattice perturbations can be achieved by further low energy stimulation. The term Charge-carrier Emission Spectroscopy can only strictly be applied when both detection and spectroscopy of negative ions, positive ions and neutrals as well as electrons are undertaken.

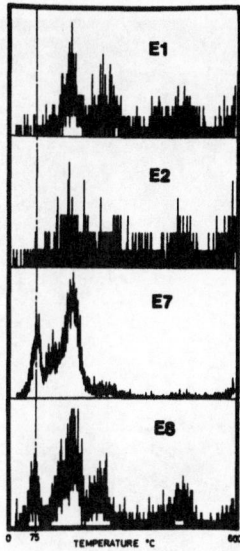

Fig. 2. Thermally stimulated exoemission curves from non-endemic
         (E1 and E2) and endemic (E7 and E8) soils. Note that the
         emission peak at 75°C is only present in endemic soils.
         From Ref (19).

infection. A link between the incidence of elephantiasis and trop-
ical red clay soils derived from volcanic basalts was suggested after
extensive epidemiological studies in Ethiopia [15] and has been supported by similar studies in Kenya, Burundi and Rwanda [16]. The disease is limited to the bare-footed sector of the population and
studies of the femoral lymph nodes of these people have demonstrated
particles within the tissue spaces of the lymph nodes and contained
within secondary lysosomes of macrophages [17]. As these particles
contain potentially irritant and fibrogenic materials the disease
has been labelled a "silicosis" of the lymphatics.

     However, the femoral lymph nodes of non-elephantiasics also
contain similar concentrations of minerals and attempts to distinguish between soils using both physical and chemical methods have
been generally unsuccessful, leading Price et al [17] to conclude
that "some physico-chemical state" is important in the aetiology
of the disease. More recently, following the use by Blanke et al [18]
of thermoluminescence (TL) to distinguish between endemic and non-
endemic soils we examined groups of both soils by TSEE following
high doses of ^{60}Co gamma-irradiation. Like TL, these measurements
proved successful in separating endemic and non-endemic samples by
the form of their emission curves on heating (Figure 2). While TL
demonstrates the presence of intrinsic lattice defects frozen into
these mixed mineral crystalline lattices during rapid cooling of

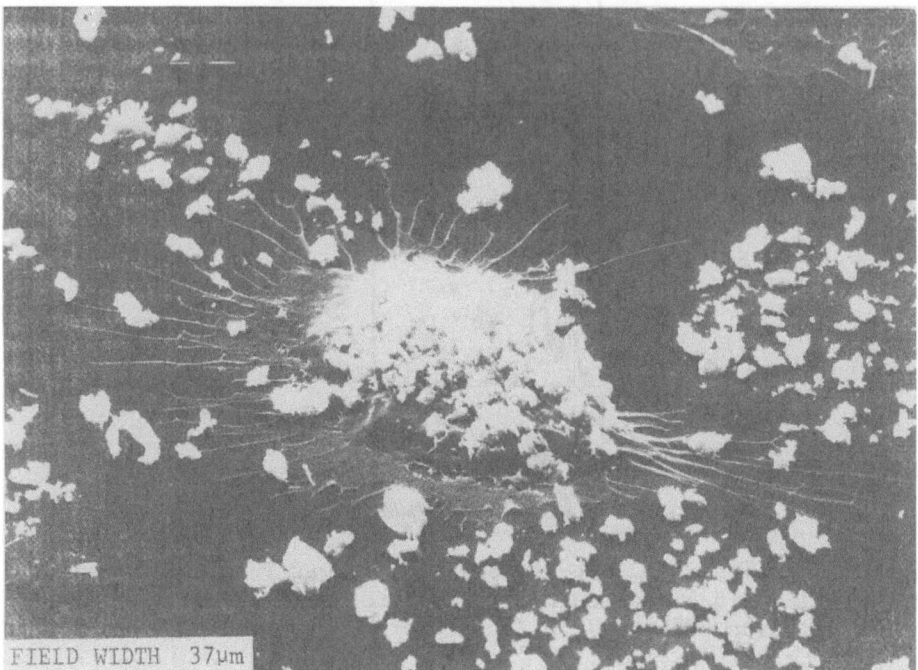

FIELD WIDTH   37µm

Fig. 3. Murine peritoneal macrophage after 6 hours culture in the
         presence of 50µgrms of sub 2-micron particles of endemic
         soil E7 per ml of culture medium. Filopodia, extending
         from the cytoplasmic skirt, are collecting particles in
         the immediate vicinity of the cell and transferring them
         to the apex of the cell body.

the volcanic material from which they are derived, exoemission demonstrated that surface reactivity of these two soil groups was also
significantly different [19]. Having established these physical methods
to obtain characteristic "fingerprints" of the materials, we under-
took a study of the cytotoxic effects of these soil particles to
macrophages in culture.

    Sub 2-micron particles of these mixed alumino-silicate dusts
were added to cultures of murine macrophages collected, by peritoneal lavage, from adult female B10 Brown mice. We combined dye-
exclusion assays, morphological assessment by both light and scanning
electron microscopy and elemental analysis of the particles to supplement our previous TSEE observations. Our initial work showed that
while there were significant differences both in Al:Si and Al:Ca
ratios ($p < 0.05$ and $p > 0.0005$ respectively) and the number of sub-
2-micron size particles in endemic with respect to non-endemic soils,

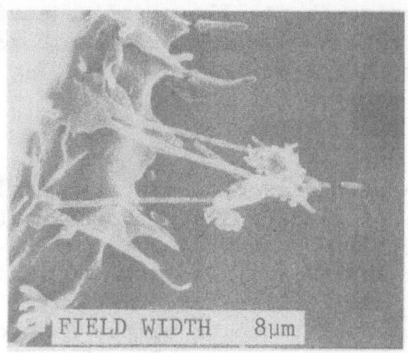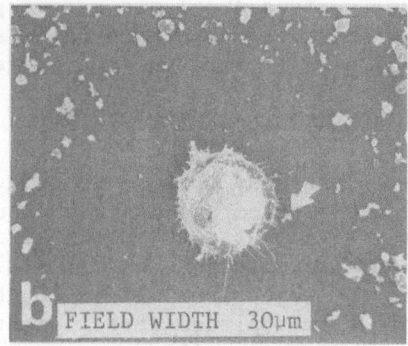

Fig. 4. (a) Four filopodia collecting an irregular shaped particle
of soil E7. Detail of area marked with arrow in (b).
(b) This cell, after a 24 hour culture period, has created
a particle free halo.

all soil samples would cause cell death following avid phagocytosis
by macrophages. However, greatest cell death was shown to occur
within the first 48 hour period indicating that some property of
the soils themselves, rather than their breakdown products, was
causing cell death and the possible dose dependence of small soil
particles was discussed [20].

These soil particles are phagocytosed in culture in a three
step process involving collection, transfer and ingestion. After 6
hours in culture (Figure 3) the majority of cells were seen to be
roughly circular. An elevated central mass containing the nucleus
and organelles is seen surrounded by a skirt of cytoplasm. From this,
filopodia extend towards the particles and gather them towards the
cell body (Figure 4(a)). The particles are transferred to the apex
of the cells where they are ingested. In this way the cells cleared
particle-free haloes around themselves which could be easily seen
after a 24 hour culture period (Figure 4(b)).

These SEM observtions dramatically illustrate the morphological
relationship between cell membranes and inorganic substrata. They
can be used, in part, to monitor the reactions of these cells in-
vitro when insulted with such toxic materials but little insight is
gained into the mechanisms responsible. However, if the dose of the
soil added to the culture medium is changed then vital-dye exclusion
tests reveal differences in the reaction of macrophages to the two
soil groups. As the additive dose of soil increases above a threshold level of 100 µgrams/ml culture to 200 µgrams/ml culture, the
endemic soils become more toxic towards the macrophages with respect
to non-endemic soils [21]. This fact, together with both the higher
number of smaller particles contained in the endemic soils (thus a

higher surface area) and a more physico-chemically reactive surface,
would indicate that the dynamic surface properties of endemic soils
are important criteria in their cytotoxicity.

BONE SUBSTITUTE CERAMICS

   In the case of bone substitute materials cytotoxicity should
not be a problem and one can assume biocompatibility. However, the
bioreactivity of such materials is still open to many questions.
For such materials to be clinically successful, two fundamentally
different objectives may be sought: the implant may be required as
a long term replacement for natural bone tissue or simply a space
maintainer which will encourage new bone formation. Such bone formation could, if the implant material be resorbed, completely replace
the artificial substrate. Ideally, these long-term bioreactive pathways should depend upon the reaction of cells to the artificial inorganic substrate.   Tricalcium phosphates, for example, are considered resorbable (biodegradable) while dense hydroxyapatite ceramics
are non-biodegradable and may even promote appositional bone growth [4].
However, the response evoked by such materials in a host tissue is
poorly understood and preparation methods and impurity levels in the
starting materials are known to effect biological reactivity at the
implant interface.   While rigorous control of implant composition
is not necessary for biocompatibility, Jarcho [22] has stated that
such control is essential to achieve predictable "in-vivo" results,
particularly with regard to bioresorption characteristics and Clarke

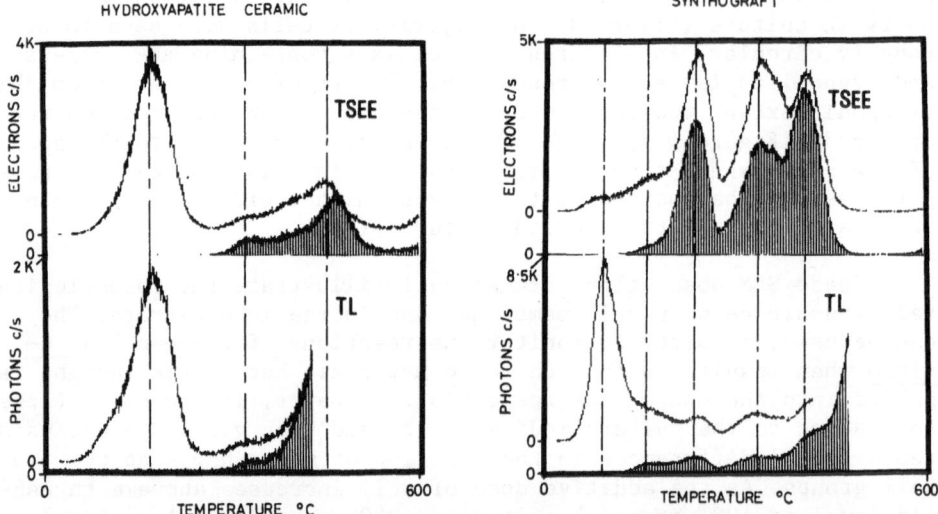

Fig. 5.   Simultaneously measured TSEE/TL from calcium phosphate
          biomaterials. Note: the hatched areas represent emission
          detected without laboratory irradiation. (From Ref (13).

et al [23] concluded that histological studies provided evidence that
ions released from implant surfaces were required for osteogenesis
although no direct evidence of ion accumulation on implant materials
was provided. More recently Klein et al [24] concluded that degradation
of calcium phosphate ceramics depends upon the crystallography and
stoichiometry of the material together with the sintering conditions
employed in production.

We have shown, with exoemission measurements, that commercially
available calcium phosphate bone-substitute ceramics exhibit significantly different electron emission "fingerprints" on thermal
stimulation [13]. Figure 5 demonstrates such results, together with
simultaneously recorded TL signals in two materials. The cross-
hatched areas represent the TSEE and TL signals of "as received"
samples which have been pulverized and sedimented onto stainless
stell discs but have received no laboratory irradiation. The solid
lines represent emission measured following ^{60}Co gamma-irradiation.
The bioresorbable material "Synthograft" (Miter Inc., USA) is clearly
more surface active (as seen by TSEE) than the hydroxyapatite ceramic.
Following laboratory irradiation the three high temperature peaks
in "Synthograft" are augmented although the total photon emissivity
(at lower temperature) is greater than that of electrons. This
suggests that while volume defects (TL) predominate following irradiation, these created perturbations must relax in time at ambient
storage temperatures (this is to be expected from the temperature
value of the main TL peak) leaving longer-term metastable states
which provide a high level of surface reactivity (high TSEE, low
TL). By comparison, in the hydroxyapatite material electron and
photon emission peaks are congruent following irradiation although
the former is more intense (4,000c/sec maximum). Neither of these
emissions is visible in the "as received" material. Furthermore,
the volume trap density as measured by the TL maxima is considerably
lower than the resorbable material. We believe that relaxation of
these types of crystalline pertubations, stored in the sample
during its preparation, will affect the behaviour of cells with
which such materials come into contact to form a biological interface.

To observe the colonization of such bone-substitute materials
by osteoblasts, we have adopted the method first described by Jones
and Boyde [25] which comprises the preparation of periosteally stripped
endocranial surfaces of neonate rat calvaria. After careful removal
of the periosteum, the surfaces of exposed parietal bones can be
seen to be covered with a confluent layer of tesselated osteoblasts
occasionally interrupted by multinucleate osteoclasts. The distribution of osteoclasts is known to change rapidly in the first seven
days of life with a drift to the periphery [26]. Culturing such preparations, Jones and Boyde [27] observed colonization, by osteoblasts,
of various calcified intact or anorganic natural substrata.

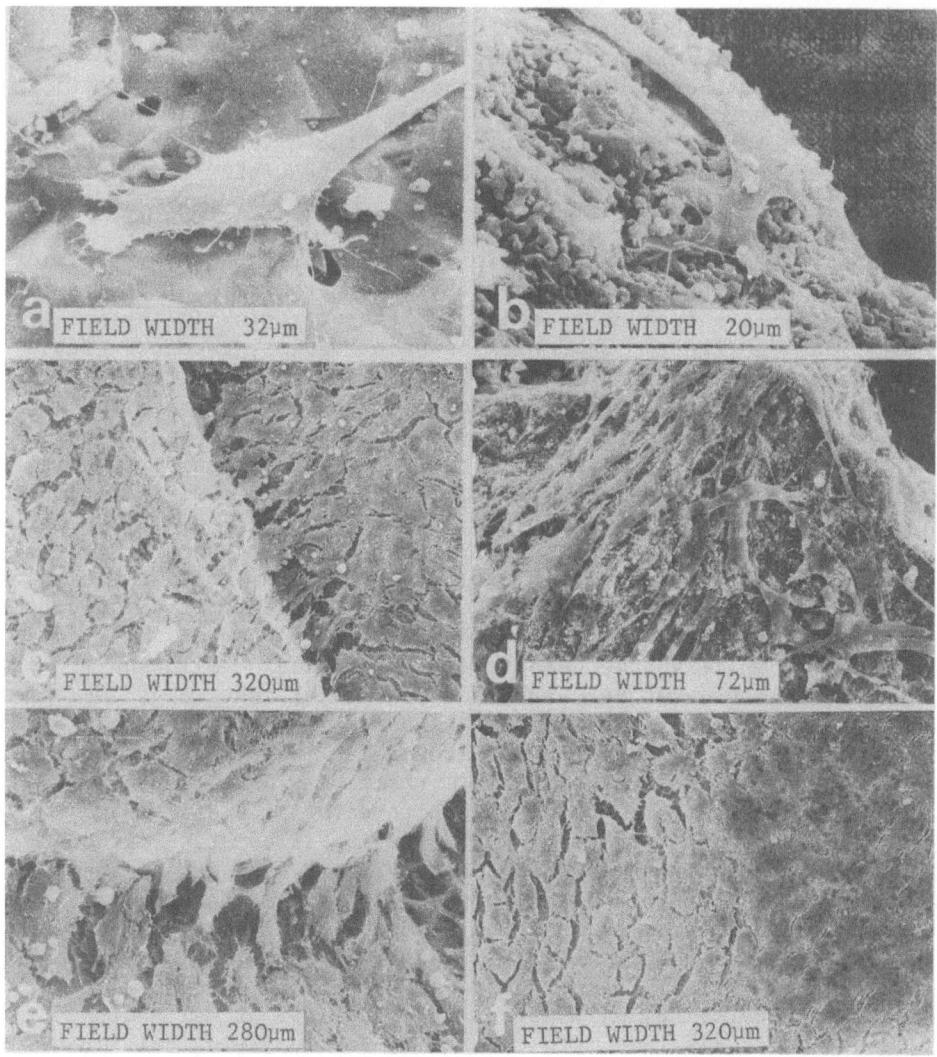

Fig. 6.   S.E.M. photomicrographs of cultured osteoblasts colonizing
bone-substitute calcium phosphate ceramics.
(a) A single murine osteoblast migrating over the surface
of "Blue Hydroxyapatite". (b) An osteoblast cell extension
on the surface of a "Synthograft" particle. (c) A confluent layer of osteoblasts migrating from the surface
of the cultured parietal bone (right) onto "Blue Hydroxyapatite" (left). (d) Numerous, considerably elongated,
cells on a "Synthograft" particle. (e) A confluent layer
of cells colonizing a sample of "Calcitite" (above).
(f) The centre of the particle shown in (e) is seen on
the right where cell density was greater than that of
migrating cells (left).

Having prepared such bone samples we transferred them to "Limbo" multiwell tissue culture plates. Each well contained 1 ml of LDA medium containing 10% foetal calf serum. A single piece of calcium phosphate ceramic material was placed on each bone fragment. The preparations, following culture periods of up to 48 hours were fixed in 2.5% phosphate buffered gluteraldehyde, dehydrated, freeze-dried and gold coated prior to viewing in the SEM (ISI Model 100A).

After 24 hours in culture large numbers of osteoblasts from rat parietal bone fragments had migrated onto all substrata. Fewer cells were observed migrating onto ceramic surfaces in similarly prepared murine cultures and thus individual cells in the leading ranks could be studied more closely. Figures 6(a) and (b) show such individual cells on "Blue Hydroxyapatite" (Phillips, Netherlands) and "Synthograft" respectively. The morphology of both individual cells and whole sheets of cells (Figures 6(c) and (d)) differed on these substrata. While we have no quantitative data, our observations suggest that osteoblasts will colonize these examples of non-biodegradable and biodegradable materials equally easily. This would concur with the observations of Jones and Boyde that osteoblasts will colonize many biocompatible substrata [25,27]. As the culture media and conditions were identical in these experiments it would seem likely, however, that the substrata is causing the differences in morphology seen in Figures 6(c) and (d). This observation is important when considering the long-term bioreactivity of such substrata since morphology is indicative of intracellular activity and hence cell metabolism. Another non-biodegradable ceramic "Calcitite" (Calcitek Inc., USA) was also facilitated examination of cell shape. Figure 6(e) shows osteoblasts colonizing this material in the same culture conditions. It was possible to observe an increase in cell density (Figure 6(f)) caused by cessation of osteoblast migration at the apex of this material. In small residual inter-cellular spaces, we found evidence of a reticular network which we interpreted as organic matrix production as has previously been reported on glass substrates [27].

In this experimental model the surfaces of examined substrata can be expected to be coated with serum proteins (present in our culture medium) which are shown to increase cell adhesiveness and facilitate cell migration. The role of the substratum, therefore, and its charge-carrier surface profile in particular, in influencing cell behaviour will be expressed through the intermediary of serum proteins and possibly other constituents of the aqueous, high dielectric-constant, environment.

GENERAL REMARKS

While simple exoemission "fingerprinting" can demonstrate subtle differences in electron emissivity of inorganic solids, measurement of other charge-carrier species emitted during relaxation of crystalline perturbations can considerably increase our understanding of surface reactivity and identify the atomic constituents of impure lattices which are responsible for their dynamic behaviour. Such measurements necessitate ultra-high vacuum equipment rather than ambient pressure detectors and have been named "Charge-carrier Emission Spectroscopy" [13]. Initial results (unpublished data) pertinent to the biological application of such techniques have demonstrated that water adsorption on apatites can increase surface emissivity and that some electron emission events are concomitant with the surface release of negative ions. We are pursuing these investigations in parallel with the development of histochemical and cell biological probes of the cells migrating onto our samples in-vitro to supplement the methods reported here in an attempt to understand behaviour at the interface.

ACKNOWLEDGEMENTS

Our inorganic substrata have been generously supplied by J.W. Frame, J.F. Kay, P.F. Lofts and E.W. Price whom we thank. We should also like to record our appreciation of the skilful help received from both Sue Dipple and Lesley Tompkins in the production of our SEM photomicrographs.

Financial support for our work from the University of Birmingham and personal support from the Jane Hodge Foundation (for R.P.H.) and a Walter Asten Scholarship (for N.T.S.) are gratefully acknowledged.

REFERENCES

1.   K. Ludloff, Untersuchungen uber den Galvanotropismus, Pfluger, Arch. Physiol. 59: 524-554 (1985).
2.   "Locomotion of Tissue Cells". Ciba Foundation Symposium, 14. Elsevier, Amsterdam (1983).
3.   see General Discussion I in Ref. (2).
4.   L.L. Hench & E.C. Ethridge, "Biomaterials" an interfacial approach", Academic Press, New York (1982).
5.   S.R. Morrison, "Electrochemistry at Semiconductor and Oxidized Metal Electrodes", Plenum Press, New York (1980).
6.   Z.A. Munir & J.P. Hirth, The nature and Role of Surface charge in Ceramics in: "Surfaces and Interfaces in ceramic and ceramic-metal systems", Materials Science Research 14, ed, J. Pask & A. Evans, Plenum Press, New York (1981).
7.   U.R. Evans, "An introduction to Metallic Corrosion" 3rd Ed, Edward Arnold, London (1981).
8.   G.C. Lowrison, Fundamentals of the Breakdown of Solids, Chapter 6 in: "Dispersion of powders in Liquids", G.D. Parfitt, ed, Applied Sci. Pub., London (1981).

9. J.E. Davies, Surface dependent emission of low energy electrons (exoemission) from apatite samples in: "Adsorption on and Surface Chemistry of Hydroxyapatite", D.N. Mistral, ed, Plenum Press, New York (1984).

10. L.A. Larson, T.A. Oda, P. Braunlich & J.T. Dickinson, Emission of Cl atoms from NaCl during V_k-centre decomposition, Solid State Communications 32: 347-351 (1979).

11. H. Glaefeke, Exoelectron emission, Chapter 5 in: "Thermally Stimulated Relaxation in solids", P. Braunlich ed, Topics in Applied Physics 37, Springer-Verlag, Berlin (1979).

12. I.V. Krylova, The chemical aspect of exoemission, Russian Chemical Reviews. Trans. from Uspekhi Khimii 45: 2138-2167 (1976).

13. J.E. Davies, Biological Applications of charge-carrier emission spectroscopy, in: Proc. 8th Polish Seminar on "Exoelectron emission and Related Phenomena", In Press: Acta. Univ. Wrat. - Mat. Fyz. Astrn.

14. J.E. Davies & P. Ramsay, A microcomputer-controlled apparatus for simultaneous measurement of exoelectron emission and thermoluminescence, Radiation Prot. Dos. 4: 177-180 (1983).

15. E.W. Price, The relationship between endemic elephantiasis of the lower legs and the local soils and climate, Trop. Geogr. Med. 26: 225-230 (1974).

16. E.W. Price, Endemic elephantiasis of the lower legs in Rwanda and Burundi, Trop. Geogr. Med. 28: 283-290 (1976).

17. E.W. Price, W.J. McHardy & F.D. Pooley, Endemic elephantiasis of the lower legs as a health hazard of barefooted agriculturalists in Cameroon, West Africa. Ann. Occ. Hyg. 24: 1-8 (1981).

18. J.H. Blanke, E.W. Price, H.M. Rendell, J. Terry, P.D. Townsend & A.G. Wintle, Correlations between elephantiasis and thermoluminescence of volcanic soil, Radiation Effects 73: 103-113 (1983).

19. J.E. Davies & P.D. Townsend, Exoemission of Ethiopian soils and the endemicity of non-filarial elephantiasis, Radiation Prot. Dosimetry 4: 185-188 (1983).

20. N.T. Spooner & J.E. Davies, The possible role of soil particles in the aetiology of non-filarial elephantiasis: a macrophage cytotoxicity assay, In Press: Trans. Roy. Soc. Trop. Med. & Hyg.

21. J.E. Davies & N.T. Spooner, Dose dependence of soil dust cyto-toxicity to macrophages, In preparation.

22. M. Jarcho, Calcium phosphate ceramics as hard tissue prosthetics, Clin. Orth. and Relat. Research 157: 259-278 (1981)

23. A.E. Clarke, L.L. Hench and H.A. Paschall, The influence of surface chemistry on implant interface histology: A theo-retical basis for implant materials selection, J. Biomed. Mater. Res. 10: 161-174 (1976).

24. C.P.A.T. Klein, A.A. Driessen & K. de Groot, Relationship
    between the degradation behaviour of calcium phosphate
    ceramics and their physical-chemical characterisation
    and ultrastructural geometry, Biomaterials. 5: 157-160
    (1984).
25. S.J. Jones & A. Boyde, The migration of osteoblasts, Cell
    Tiss. Res. 184: 179-193 (1977).
26. N.A. Barnicot, The supravital staining of osteoclasts with
    neutral red, their distribution on the parietal bone of
    normal growing mice and a comparison with the mutants
    grey-lethal and hydrocephalus-3, Proc. Roy. Soc. (Lond)
    134: 467-485 (1947).
27. S.J. Jones & A. Boyde, Colonization of various natural sub-
    strates by osteoblasts in-vitro, Scanning Elect. Microscopy
    1979/ii: 529-538 (1979).

SECTION 3

CELLULAR INTERACTIONS IN VASCULAR GRAFTS

AND EXTRACORPOREAL CIRCUITRY

CELLULAR INTERACTIONS IN EXTRACORPOREAL CIRCUITRY

Patricia Lawford

Department of Medical Physics & Clinical Engineering
Royal Hallamshire Hospital
Sheffield S10 2JF

INTRODUCTION

A major problem in the development of all extracorporeal circuitry where blood is in contact with synthetic materials is the number
and diversity of undesirable interactions may involve alterations in
both blood constituents and the synthetic materials themselves. For
example, in surgical practice, blood trauma is often a limiting factor
in the duration of ex-vivo procedures and may also cause significantly
abnormal physiological states during chronic procedures such as haemodialysis.

This paper gives a brief overview of the current understanding
of how and why blood components are traumatised or interact during
circulation through extracorporeal devices and emphasizes the importance of a multidisciplinary approach to circuit design.

EXTRACORPOREAL DEVICES

Extracorporeal devices have been in use since the late 1930's
to replace or augment natural processes. The two types of systems
most widely employed are blood purifiers and blood oxygenators.

1. Blood Purifiers

Several methods of blood purification are available for clinical
use, for example, haemodialysis, peritoneal dialysis, haemoperfusion,
haemofiltration and plasmapheresis. Of these, haemodialysis remains
the most commonly employed technique. During this procedure waste
products are removed from the blood in a membrane separation device,
the dialyser.

111

Dialysers may be described as a coil, flat plate or hollow fibre type, based on the design configuration of the membrane. In each case blood and dialysate are separated by a semipermeable membrane which is usually made of cellophane, cuprophane or cellulose acetate.

The coil dialyser consists of a cellophane tube supported on a nylon mesh, wound round a central core through which blood is pumped. In the flat plate dialyser blood flows within a membrane envelope formed by two sheets of polymer. The hollow fibre dialyser consists of a large number (5,000-20,000) of capillary tubes arranged in parallel through which the blood flows. Dialysate flows over the outside of the fibres in a counter-current configuration, promoting an increased efficiency of clearance compared with like-current flow.

2.   Oxygenators

Blood oxygenators fall into two categories; bubble or membrane oxygenators. Bubble oxygenators have the advantage of efficient and rapid transfer of gases to and from the blood but invariably this is at the expense of some degree of blood cell trauma. Blood is pumped into a container in direct contact with fine bubbles of carbon dioxide and oxygen gas. This results in considerable foaming of the blood which must then be pumped through a defoaming unit consisting of mesh treated with a defoaming agent. Finally, the blood passes through a "bubble trap" to remove any remaining air bubbles before being returned to the patient. The high interfacial energy between the blood and gas phase may cause haemolysis, red cell, platelet and white cell damage.

In the membrane oxygenator, a diffusion membrane separates the oxygen from the blood in a counter current flow system. As there is no direct contact between blood and gas, this system is potentially less traumatic to blood components. Membrane oxygenators are, however more expensive, more complicated to use and require more intensive intraoperative monitoring then the bubble type.

EXTRACORPOREAL SYSTEMS

Inspite of their different functions, the circuits in which these devices are employed are similar in outline design and material problems. They consist of a series of tubes and cannulae transfering blood from the patient to the oxygenator or dialyser, connected to one or more pump system to circulate the blood and heat exchanger to maintain a constant temperature (Figure 1).

MATERIALS

All materials used in these circuits must be non-toxic, of low thrombogenecity and mechanically durable. In addition, membranes must have the correct properties for efficient gas or solute ex-

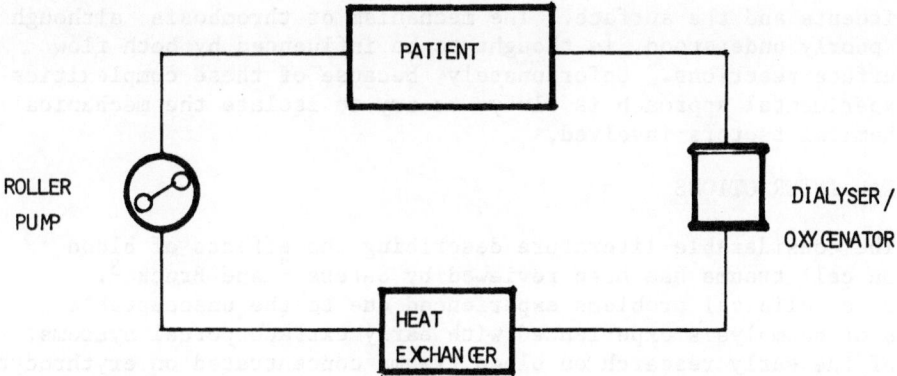

FIGURE 1.   A basic outline of the extracorporeal circuit

change.   Early membranes were made from cellulose or cellulose-based
derivatives.   More recently, other polymers have been used.   Many
other mad-made materials are employed for tubing, connectors and
reservoirs of which just a few more common examples are given in
Table 1.

CELL-CIRCUIT INTERACTIONS

It can be seen that during its passage through the extracorporeal
circuit the blood is exposed to large areas of man-made materials.
This may result in blood cell consumption due to haemolysis, thrombosis and cell adhesion leading to thrombocytopenia, leucopenia, loss
of white cell function against infection and risk of thromboembolism
within the patient's circulation.   Problems of blood coagulation are
minimized by systemic administration of heparin.   This increases the
risk of haemorrhage and is unable to prevent thrombus formation due
to platelet aggregation which has been shown to occur in spite of
adequate heparin administration [1].   The immediate goal of extracorporeal perfusion technology is to prevent thrombosis within the
circuit whilst maintaining a nominal degree of coagulation in surgical
wounds.

The problem of blood cell damage is not purely one of blood
compatibility with artificial surfaces or of flow-induced mechanical
damage.   Mechanical and chemical processes may act in a synergistic
fashion.   For example, mechanical interactions between blood cells
and a foreign surface in addition to causing physical damage to the
cells may also allow chemical reactions to take place between membrane

constituents and the surface. The mechanism of thrombosis, although still poorly understood, is thought to be influenced by both flow and surface reactions. Unfortunately, because of these complexities the experimental approach is always to try to isolate the mechanical and chemical factors involved.

CELLULAR INTERACTIONS

The considerable literature describing the effects of blood flow on cell trauma has been reviewed by Sutera [2] and Bruck [3]. Because of clinical problems experienced due to the unacceptable levels of haemolysis experienced with early extracorporeal systems, much of the early research on blood trauma concentrated on erythrocyte damage. Platelets have attracted attention more recently whilst the effects on leucocytes are more poorly understood.

TABLE 1

MATERIALS USED IN EXTRACORPOREAL CIRCUITS

MEMBRANES

Cellulose Based

Cupruphan
Cellulose Acetate (Celanese)

Other

Polymethylmethacrylate
Polyacrylonitrile
Ethylcellulose Perfluoresters

TUBING

Silicone rubber
Polyvinyl chloride
Teflon
Segmented polyrethanes

1.   Erythrocytes

A considerable number of in vitro experiments have been performed to investigate the effects of shear-stress on red cells. Whenever blood flows over a solid surface a shearing action occurs in the blood adjacent to the surface. If blood is pumped through a tube (Figure 2) the velocity will be at a maximum at the centre of the tube decreasing gradually to zero at the wall. Shear forces

are greatest at the wall where the slow motion of the blood resists
the motion of the faster flowing core. Such forces acting on the
blood cell membrane may cause stretching or even cell lysis.

The intensity of shear stress causing red cell lysis in vitro
is open to debate and the results given by several groups of workers
vary by several orders of magnitude from 100 to 4,000 N/m [2,4]. As
a variety of experimental techniques have been employed to obtain
this data, one factor which must be considered is the degree to
which the containing surfaces might be responsible for the haemolysis observed [5,6]. Haemolysis occurring as a result of red cell
contact with the containing vessel cannot be distinguished from
that caused by shear stress. The duration as well as the intensity
of shear stress applied must also be taken into account as cells
have been shown to tolerate extreme levels of shear for short periods
of time.

Haemolysis may be due to factors other than high shear stress
associated with laminar flow. Prolonged turbulent flow which is
likely to occur in regions of flow separation for example as blood
enters a manifold it may lead to red cell deformation and subsequent
haemolysis when cells are exposed to further shearing stresses within
the circuit. Sudden negative pressures due to cavitation or circuit
failure, for example, undetected arterial line collapse when pumping
is allowed to continue, result in a snap explosive release of intracellular oxygen out of solution and subsequent cell lysis.

Although some cells are damaged irreversibly others (probably
a larger number) undergo sublethal trauma followed by splenic removal
in the ensuing days. Sublethal damage is characterised by post-
perfusion anaemia caused by decreasing cell life spans due to splenic
removal and autohaemolysis. Loss of membrane lipids and elevated
plasma levels of haemoglobin and lactic dehydrogenase in the absence
of overt lysis suggest that red cell fragmentation may occur with
partial loss of these intracellular constituents and subsequent re-
sealing of the gragments. It has been suggested by Indeglia and
Bernstein [7] that these resulting fragments are more spherical and
less deformable than the parent cell and therefore more susceptable
to osmotic stress. Reports by Piroski [8] that antibodies could be
detected on the surface of red cells after passage through extracorporeal circulation have not been substantiated.

In situations of low shear it is possible for red cells to
spontaneously aggregate to form costs or rouleaux. This process
is reversible and rouleaux are redispersed when external forces
increase. Although it is not known to what extent this may occur
in extracorporeal systems it is unlikely to have any clinical
significance as further aggregation of cells into larger aggregates
bound by fibrin is prevented by routine heparin administration.

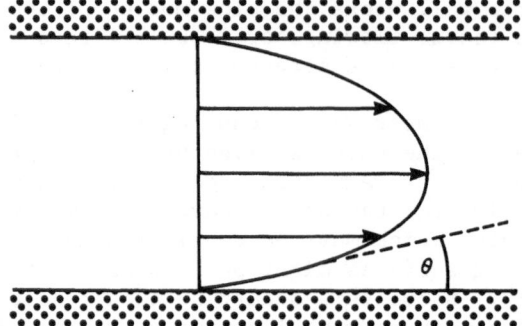

FIGURE 2.   Laminar Flow pattern: the velocity profile of blood
            flowing through a uniform tube.   The boundary layer
            adjacent to the wall is the region of most intensive
            shearing action.   The smaller the angle the greater
            the shear stress.

2.   Leucocytes

Leucocytes undergo marked morphological and metabolic changes
after interaction with foreign surfaces.   Haemodialysis, for example,
is associated with a marked but transient and rapidly reversible
fall in the patient's leucocyte count.   This phenomena is due to a
decrease in neutrophil and monocyte counts and has been shown to be
membrane dependant, the degree of leucopenia observed being more
profound after blood contact with cellulose-based membranes [9].

This decrease in leucocyte count coupled with a decrease in the
observed phagocytic activity of surviving leucocytes [10] suggests a
potential impairment of the patient's capacity to resist infection.
However, the clinical significance and mechanism of leucocyte-
membrane interactions have yet to be fully established [11].

Artificial membranes are thought to cause activation of the
complement system by the alternative pathway [12].   Circulating
complement reacts with neutrophils causing modifications in cell
shape and stimulating cell aggregation.   Craddock et al [13,14]
suggested that such circulating aggregates may become trapped in
the lung microvasculature giving rise to pulmonary dysfunction and
the hypoxia often associated with haemodialysis.   This hypothesis
has met with some degree of criticism [15,16].

Leucocytes may suffer damage from shear stress at levels ten

times less than those affecting normal erythrocytes. As these
stresses are just above those said to be encountered physiologically,
any flow perturbations within the extracorporeal circuit which lead
to increased shear force are likely to stimulate leucocyte aggregation.

Platelets

The effect of blood circulation through extracorporeal circuitary
on platelet count and volume is shown from recent work performed at
Sheffield University by Martin et al [17]. A profound change in
platelet count and volume was shown to occur in patients during the
first forty days following the use of a cardiopulmonary bypass
system incorporating a bubble oxygenator during surgery for coronary
artery or heart valve replacement. In the immediate post operative
period mean platelet count fell significantly to approximately 60%
of the preoperative level with no change in mean platelet volume,
indicating that platelet loss occurred with no preferential removal
of large platelets. This period of thrombocytopenia was sustained
for three days during which time there was a significant increase
in mean platelet volume followed by a period of increasing platelet
count to give a rebound thrombocytosis of approximately 250% with
a significant decrease in mean platelet volume Figures 3 a and b.

These changes in platelet count and volume after thrombocytopenia are similar in pattern to those observed by the same group
of workers after platelet depletion induced in rabbits by injection
of antiplatelet serum although the quantitative response is not as
great [18,19], Figure 4.

Table 2 compares the count, volume and function of platelets
in this animal study before and after 24 hours of antiserum induced
thrombocytopenia. The results show a significant increase in
platelet count and volume but a decrease in the volume of circulating platelet cytoplasma per litre of blood. Production of thromboxane A_2, a powerful platelet aggregating agent and vasoconstrictor,
by the platelets was taken as an index of haemostatic potential.

Platelets were subjected to three different stimuli; sodium
arachidonate, collagen and thrombin. In each case thromboxane
production per 10 million platelets increased after 24 hours of
thrombocytopenia. Thromboxane production per litre of blood remained constant due to a significant increase in thromboxane
produced per unit volume of cytoplasm. These animal studies show
that the larger platelets produced after thrombocytopenia are more
haemostatic than control platelets. The authors suggest that the
large haemostatic platelets produced soon after bypass followed by
a substantial thrombocytosis might be involved in the cerebro-
vascular disturbances occasionally reported after the use of this
ex-vivo technique.

	CONTROL n = 12			AFTER 24 HOURS OF THROMBOCYTOPENIA n = 5		
Platelet count * (x 10^{-9} (pl/ l blood))	455.9 : (SEM = 158.7)			150.5 : (SEM = 27.72) †		
Platelet mean volume (fl.)	5.92 : (SEM = 0.39)			7.25 : (SEM = 0.59) †		
Volume of platelets / l blood (x 10^4 fl. pl / l blood)	26.98			10.91 †		
	1.5 mM NaAA	4 µg/ml collagen	1 u/ml thrombin	1.5 mM NaAA	4 µg/ml collagen	1 u/ml thrombin
Thromboxane produced (p mol / 10^7 platelet)	4.52 (SEM = 0.52)	1.26 (SEM = 0.33)	1.48 (SEM = 0.54)	14.25 † (SEM = 4.97)	3.66 † (SEM = 1.45)	4.83 † (SEM = 0.67)
Thromboxane production/ l blood (x 10^4 p mol/ l blood)	20.60	5.74	6.75	21.45	5.50	7.27
Thromboxane production per unit volume of platelet (x 10^7 p mol/fl. pl)	0.76	0.21	0.25	1.97 †	0.50 †	0.67 †

* Platelets/ litre of blood   † Significant difference

TABLE 2. Differences in platelet count, volume and function in rabbits following antiserum–induced platelet depletion. Platelets were stimulated by exposure to sodium arachidonic acid, collagen or thrombin and the resulting thromboxane A_2 production was taken as an index of haemostatic potential.

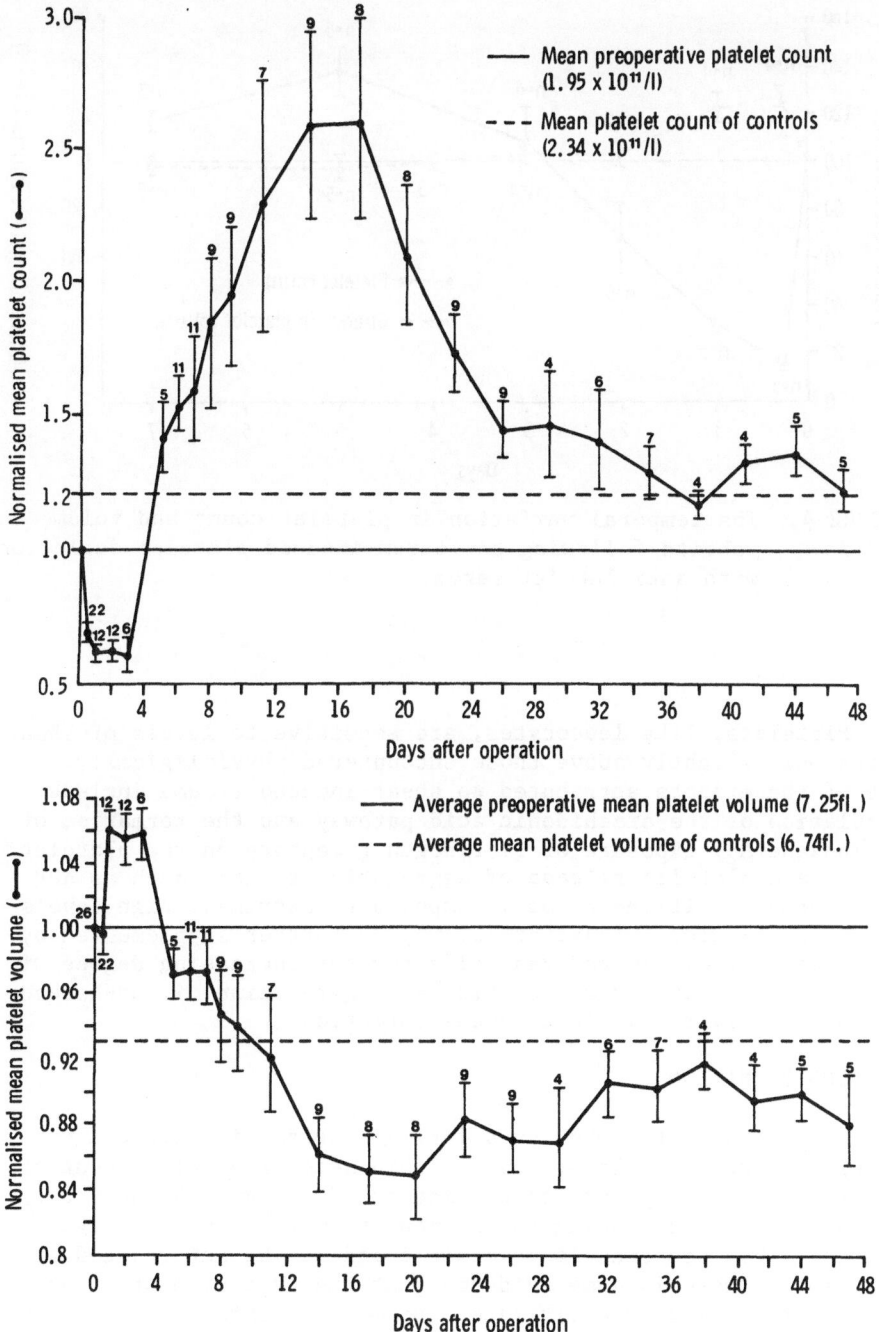

FIGURE 3 a and b. The temporal variation of mean platelet count
and mean platelet volume after cardiopulmonary bypass during heart
surgery. Both count and volume were normalized with respect to
their respective pre-operative levels. The bars represent one
standard error of the mean for the number of patients sampled.

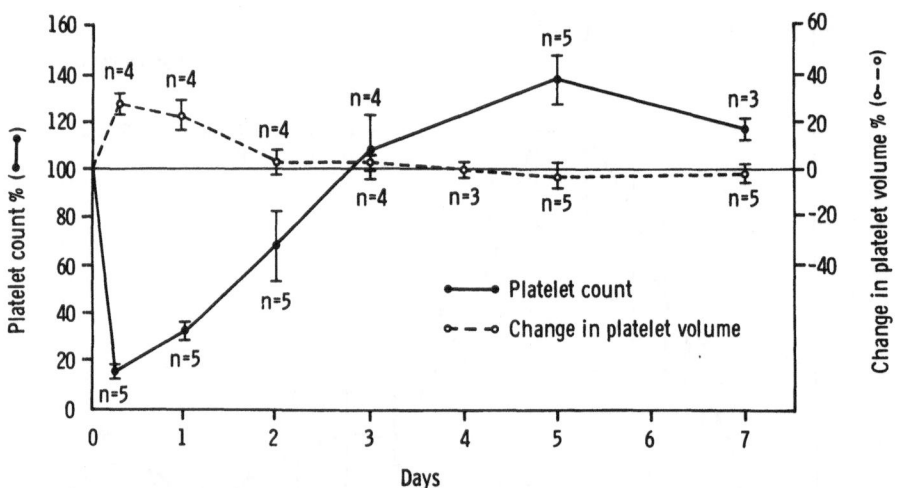

FIGURE 4. The temporal variation in platelet count and volume in
rabbits following antiserum-induced platelet depletion
with antiplatelet serum.

Platelets, like leucocytes, are sensitive to levels of shear
stress only slightly above those encountered physiologically.
Some of the effects attributed to shear-induced trauma include
stimulation of the arachidonic acid pathway and the formation of
thromboxane A_2, exposure of fibrinogen receptors on the platelet
surface and platelet release of aggregation factors such as ADP
and serotonin. If the blood is exposed to extremely high levels
of stress platelet release of ADP may be further supplemented by
ADP released from damaged red cells further increasing degree of
aggregation. Platelet stimulation and aggregation, if unchecked,
will ultimately result in thrombus formation.

THROMBUS FORMATION

Formation of thrombus on a foreign surface is preceded by
protein absorption. This occurs within seconds of blood contact
and is influenced by surface parameters [20]. Figure 5 shows the
sequence of events leading to thrombus formation on a foreign
surface. This sequence is reversible and may be interrupted at
any point. There is some evidence that the type of protein adsorbed
has an influence on the ultimate outcome of events. This, in turn,
depends on the surface properties of the materials involved. Preferential coating of the foreign surface with albumen, for example,
prevents platelet adhesion. This technique has been used experimentally in an attempt to obtain a renewable antithrombotic surface

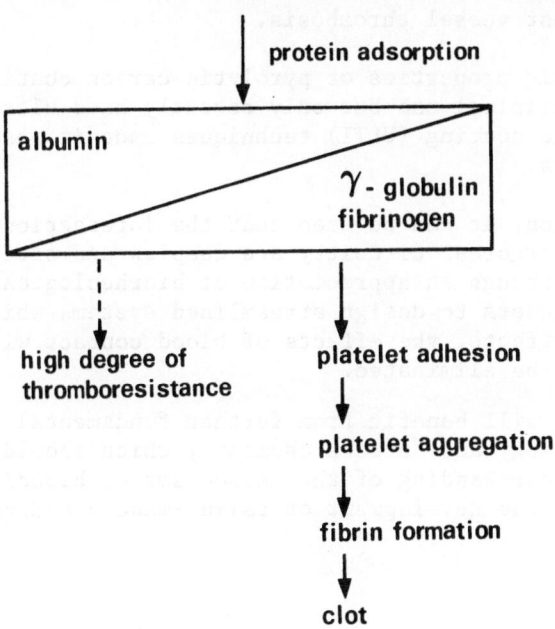

FIGURE 5.   The sequence of events at a blood-polymer interface
            leading to the formation of thrombus.

but unfortunately, only proved effective when surface area to
blood volume ratios are small.

    Haemodynamic factors may also play a direct part in preventing
thrombus formation. High flow rates resulting in high surface shear
forces may detach blood elements from the artificial surface before
thrombus is formed, reduce contact time preventing cell/material
interactions or may remove formed thrombus causing embolisation
leaving a clean surface. Conversely, it can be argued that at high
flow rates the rate at which platelets are brought to the surface
increases. If they were able to adhere a more rapid accumulation
of platelets may result.

    Recent methods of controlling adverse blood/material interactions
have been directed at breaking this chain of events leading to
thrombus formation. Several methods have been proposed. Antithrombotic coatings, covalently linked heparin, for example, have
been used successfully to prevent coagulation within the circuit [29].
The degree of success is increased considerably when coupled with
systemic administration of an anti-platelet agent such as prostacyclin. Sugitachi et al [22] coated PVC and silicon rubber with

plasminogen activators in order to stimulate fibrinolysis by activation of the plasmin system thus stimulating the normal in vivo protection against vessel thrombosis.

The antithrombotic properties of pyrolytic carbon coatings are well established in clinical use but only recently have ultra low temperature isotropic coating (ULTI) techniques made it possible to coat flexible surfaces.

In conclusion, it can be seen that the interactions between cells and extracorporeal circuitry are complex and often result in cell trauma. Although an appreciation of biorheological factors has enabled engineers to design streamlined systems which minimise shear and flow effects, the effects of blood contact with man-made materials cannot be eliminated.

The subject will benefit from further fundamental research in terms of haematology and surface chemistry which should lead to both a fuller understanding of the mechanisms of blood/material interactions and the development of tailor-made blood/compatible biomaterials.

REFERENCES

1.   P.R. Keshaviah, D.A. Luehmann, Risks and hazards associated
        with dialysers and dialysated delivery systems. Crit.Rev.
        in Biomed. Eng., 9:201 (1983).
2.   S.P. Surtera. Flow induced trauma to blood cells. Circ. Res.
        41:2, (1977).
3.   P.L. Blackshear. Mechanical hemolysis in flowing blood. In:
        Biomechanics: Its Foundation and Objectives. Edited by
        Y.C. Fung Prentice-Hall Englewood Cliffs (1972).
4.   S.D. Bruck. Properties of Biomaterials in the Physiological
        Environment. CRC Press, Boca Raton (1980).
5.   J.A. Rooney. Hemolysis near an ultrasonically pulsating gas
        bubble. Science, 16:869, (1970).
6.   A.R. Williams, D.E. Hughes and W.L. Nyborg. Hemolysis near a
        transversely oscillating wire. Science 169:871.(1970).
7.   R.A. Indeglia and E.F. Bernstein. Selective lipid loss
        following mechanical erythrocyte damage. Trans.Am.Soc.
        Artif.Intern Organs 16:37. (1970).
8.   B. Pirofsky. Hemolytic anemia complicating aortic valve
        surgery - A new autoimmune syndrome. Blood, 24:839, (1964).
9.   N.A. Hoenich, C. Woffindin, M. Qureshi and D.N.S. Kerr.
        Membrane-induced leuocopenia. Contr.Nephrol. 36:1, (1983).
10.   B. Kusserow, R. Larrow and J. Nichols. Perfusion and surface-
        induced injury to leuocytes. Fed. Proc. 30, 1516, (1971).
11.   L.S. Kaplow and J.A. Goffinet. Profound neutropenia during
        the early phase of hemodialysis. J.Am. Med. Ass. 208:133
        (1968).

12. P. Aljama, P.A.E. Bird, M.K. Ward, T.G. Feest, W. Walker, H. Tanboga, M. Sussman and D.N.S. Kerr. Hemodialysis-induced leucopenia and activation of complement: effects of different membranes. Proc. Eur. Dial.Transplant.Ass. 15: 144, (1978).

13. P.R. Craddock, J. Fehr, A.P. Dalmasso, K.L.Brigham and H.S. Jacob. Hemodialysis leukopenia. Pulmonary vascular leuko-stasis resulting from complement activation by dialyser cellophane membranes. J.Clin.Invest. 59:879, (1977).

14. P.R. Craddock, J. Fehr, K.L. Brigham, R.S. Kronenberg and H.S. Jacob. Complement and leukocyte-mediated pulmonary dys-function in hemodialysis. New Engl.J.Med. 296:769, (1977).

15. N.W. Aurigemma, N.T. Feldman, M. Gottlieb, R.H. Ingram, J.M Lazarus and E.G. Lowrie. Arterial oxygenation during haemodialysis. New Engl.J.Med. 297:871, (1977).

16. J.E. Sherlock, J.W. Ledwith and J.M. Letteri. Hypoxemia during dialysis. New.Engl.J.Med. 297:558, (1977).

17. J.F. Martin, T.D. Daniel and E.A. Trowbridge. Acute and chronic changes in platelet count and volume after cardiopulmonary bypass. J.Card.Thor.Surg. (submitted for publication 1985).

18. E.A. Trowbridge and J.F. Martin. An analysis of the platelet and polyploid megakaryocyte response to acute thrombocyto-penia and its biological implications. Clin. Phys. Physiol. Meas. 5: 263, (1984).

19. J.F. Martin, E.A. Trowbridge, G. Salmen and J. Plumb. The biological significance of platelet volume: Its relationship to bleeding time, platelet thromboxane B_2 production and megakaryocyte nuclear DNA concentration. Throm.Res. 32: 443, (1983).

20. S.L. Cooper, B.R. Young and M.D. Lelah. The physics and chemistry of protein-surface interactions in: Interactions of the Blood with Natural and Artificial Surfaces. Edited by E.W. Salzman, Marcel Dekker Inc. New York. (1980).

21. G. Schemer, L.N.L. Teng, J.E. Vizzo, U. Graefe, J. Militunovich, J. Cole, B.H. Scribner. Clinical use of a totally heparin grafted hemodialysis system in uremic patients. Trans. Am. Soc. Artif.Intern Organs. 23: 177 (1977).

22. A. Sugitachi, M. Tanaka, T. Kawahasa, N. Kitamura and K. Takagi. A new type of drain tube, Proc. of the 3rd Ann. Meeting of the Inst. Soc. for Artif. Organs, Paris, (1981).

INTERACTIONS OF WHITE CELLS AND THE HUMORAL CASCADES DURING CARDIOPULMONARY BYPASS IN MAN

Stephen Westaby

Department of Surgery
Cardiothoracic Division
Hammersmith Hospital and Royal Postgraduate Med.School
London

For the past 30 years cardiopulmonary bypass has been used routinely for accurate repair of congenital and acquired cardiac defects. During this time the morbidity associated with both the surgical procedure and extracorporeal circulation itself have dramatically decreased and for some types of operation, such as coronary artery bypass grafting, the mortality approaches zero. However, damaging effects from cardiopulmonary bypass itself, still occur in every patient and especially in the very young, elderly or very sick patient may cause death, even after a successful cardiac repair.

The classical 'post perfusion syndrome' frequently seen in florid form 25 years ago was usually comprised of non cardiogenic (and often haemorrhagic) pulmonary oedema, renal failure, bleeding tendencies and pyrexia of non infective origin [2-6]. Cerebral dysfunction, in part related to hypoxia was also common [7]. Such changes were attributed to exposure of blood to abnormal events in the extracorporeal circuit. These can be broadly split into three groups: exposure to shear stress, incorporation of abnormal substances and exposure of blood to unphysiological surfaces.

Cellular damage due to shear stress in the roller pumps and in the oxygenator itself mainly affects red cells causing haemolysis and to some extent destroys older white cells. Better roller pumps, haemodilution and improved oxygenator technology have substantially decreased these problems.

Incorporation of abnormal substances such as fat globules, fibrin, gas bubbles and tissue debris produce microemboli which for many years were regarded as the principle aetiological factor

125

in bypass related organ damage. Again, better filters have improved
this situation though the post perfusion syndrome persists in a mod-
erated form.

The effects of exposure of blood to unphysiological, but all-
egedly 'biocompatible', surfaces has been poorly understood since bypass began. Platelets adhere to foreign materials, change their configuration and initiate thrombosis so that without heparinisation
bypass could not continue [8.9]. Large protein molecules can be de-
natured at a foreign interface, either solid or gaseous in form, and
activation of Hageman factor initiates the clotting cascade[10.11].
However, until recently the other humoral cascades, complement, fibrinolysis and kallikrein have been largely ignored. For some years
it has been known that there are close interrelationships between
these systems and that without control mechanisms of protease inhibitors, all four cascades could be activated with dramatic consequences. In fact such uninhibited intravascular activation can lead to
massive anaphylatoxin and histamine release with a fatal outcome [12].

As the mortality of intracardiac repair becomes smaller the importance of neutralising the damaging effects of cardiopulmonary bypass increases. In 1980, at the Cardiac Surgical Unit of the University of Alabama, a unifying hypothesis for the complex cellular and
humoral events was evolved. Certain features of the post perfusion
syndrome, namely interstitial oedema, fever and leucocytosis resemble
the acute inflammatory response normally limited to a site of local
injury. Could the systemic changes after bypass, represent a 'whole
body inflammatory response' to the foreign materials of the bypass
circuit? Since the body reacts so effectively to every other intrus-
ion, it is naive to expect to expose our total blood volume many
times over to a host of foreign surfaces without inflicting some
sort of defensive reaction. If it is a 'systemic' inflammatory response what is the likely mediator?

In 1972 Parker and others identified a significant decrease in
serum complement after bypass which fitted well with the findings of
the others that the post operative immune response to bacteria is
altered [13.14]. At that time it was impossible to determine whether
complement consumption was due to denaturing or triggering of the
complement cascade. If the latter, then does bypass initiate release
of powerful anaphylatoxins? [2.15]. Shortly afterward haemodialysis
and leukaphoresis workers identified systemic leukopenia, pulmonary
white cells sequestration and transient pulmonary dysfunction in
their patients on extracorporeal circuits, and proposed alterations
in neutrophil mobility secondary to C_5a anaphylatoxin release [16.17].
Similar findings were noted during cardiopulmonary bypass and used
by Hammerschmidt and coworkers to imply complement activation [18].
However, their data for total complement levels and whole C_3 and C_5
fragments did not support this hypothesis or clarify the relationship between complement activation and peripheral leukopenia. Also

Cooper had previously shown that infusion of activated complement
into sheep produced a rise in serum concentrations of prostaglandin
metabolites (thromboxane B_2) and that treatment with prostaglandin
inhibitors (indomethacin or sulphinpyrazone) eliminated both the
prostaglandin rise and pulmonary dysfuction without altering the
peripheral leukopenia [19].

At this time the cellular immunology group and the Scripps Institute, California isolated, synthesised, and developed an immunoassay
for the potent anaphylatoxins C3a and C5a [20]. Both are liberated by
the classical and alternative pathways of activation and have physiological properties including increase in capillary membrane permeability, smooth muscle contraction, histamine release from mast cells
and polymorphonuclear chemotaxis [15]. Chenoweth with coworkers at the
University of Alabama showed, in adult patients undergoing coronary
artery surgery, that C3a appeared rapidly in the serum at the onset
of bypass and accumulated in increasing quantity until the end of
perfusion when levels fell rapidly due to tissue binding [21]. Differential white cell counts from the right and left atria during perfusion, and after re-establishing pulmonary blood flow at partial bypass, showed trapping of neutrophils and monocytes in the pulmonary
vascular bed. Lymphocytes and eosinophils which do not have binding
sites for C5a, were unaffected. Estimation of C4 levels showed these
to be unchanged so activation had occurred by the alternative pathway. Subsequent investigations showed the anion-cation reaction between heparin and protamine also to activate complement but in this
case by the classical route [22]. Practically all the foreign materials
of the bypass circuit are capable of complement activation in vitro
and the rate of anaphylatoxin release is dependent on temperature
kinetics. Nylon tricot used in both oxgenators and cardiotomy res-
evoirs is a particularly potent activator [1].

In an attempt to identify possible relationships between the
quantity of anaphylatoxin released and the postoperative morbidity
a wide variety of patients with both congenital and acquired heart
disease were studied prospectively by Westaby, Kirklin and others [23].
The level of C3a three hours after bypass was of greatest predictive
value and could be related to pulmonary and renal dysfunction, also
postoperative bleeding. Cardiac output was inversely related to C3a
level at this time.

For infants higher C3a levels were found in those undergoing
a period of total circulatory arrest which probably results from
prolonged static exposure of blood to the materials in the pump oxygenator. The association between C3a level and organ dysfunction
suggests that complement activation may well play a role in producing
a 'whole body inflammatory response'.

It is probably that anaphylatoxin production acts only as a trigger for subsequent events. At the Hammersmith Hospital we have recent

quantified white cell trapping and have shown this to occur principally at the time of aortic cross clamp release, when circulation is
restored to the bypassed and consequently ischaemic lungs. As many
as 50% of neutrophils may be taken out of circulation at this time
and membrane damage is thought to occur after degrannulation and
fragmentation of these cells with release of lysosomal enzymes and
generation of oxygen free radicals [23,24]. We have demonstrated release of polymorphoneuclear elastase after cross clamp release and
have implied the presence of free radical activity and lipid per-
oxidation by measuring a simultaneous increase in malon dialdehyde
levels, coinciding with white cell trapping. During bypass the
morphology of the white cells changes and through electron microscopy
they can be seen in intimate contact with the pulmonary capillary
membrane (Westaby S., Fleming J., Royston D., Krauss T - unpublished
data).

As in other forms of the adult respiratory distress syndrome the
pulmonary changes were characterised by increased permeability of
the capillary endothelium. Subsequently widening of the alveolar
surface tension and atelectasis. This sequence of events probably
occurs in every patient after cardiopulmonary bypass but in small
infants, or the elderly and debilitated, leads to prolonged intu-
bation, increased susceptibility to infection and a greater like-
lihood of pulmonary death through inability to compensate for them.

Surprisingly,it has taken 25 years to formulate a unifying
hypothesis for the complex cellular and humoral events which occur
during cardiopulmonary bypass. With an understanding of the central
role of complement and white cells it is a simple matter to identify
areas for therapeutic intervention. However, there are many other
contributors such as platelet factors, kinins, fibrinopeptides and
products of arachidonic acid metabolism whose more peripheral but
synergist effects are poorly defined. Considerably more research is
therefore required before the damaging effects of extracorporeal
circulation can be completely eradicated.

REFERENCES

1. S. Westaby, Complement and the damaging effects of cardiopulmon-
       ary bypass, Thorax 38: 321-32 (1983).
2. N.B. Ratliff, W.G. Young, D.B. Hackel, E. Mikat, and J.W. Wilson,
       Pulmonary injury secondary to extracorporeal circulation,
       J. Thorac. Cardiovasc. Surg. 75: 104-120 (1978).
3. A.S. Geha, A.D. Sessler, and J.W. Kirklin, Alveolar-arterial
       oxygen gradients after open cardiac surgery, J. Thorac.
       Cardiovasc. Surg. 51: 609-615 (1966).
4. P. Gailiunas, R. Chawla, J.M. Lazarus, L. Cohn, J. Sanders,
       and J.P. Merrill, Acute renal failure following cardiac
       operations, J. Thorac. Cardiovasc. Surg. 79: 241-243 (1980).

5.   C.J. Lambert, A.J. Marengo-Rowe, J.E. Levenson, R.H. Green,
     J.P. Thiele, G.F. Geisler, M. Adam, and B.F. Mitchell, The
     treatment of postperfusion bleeding using E-aminocaprioc
     acid, cryoprecipitate, fresh frozen plasma and protamine
     sulphate, Ann Thorac. Surg. 28 440-444 (1979).
6.   F.B. Livelli, R.A. Johnson, and M.T. McEnany, Unexplained
     in-hospital fever following cardiac surgery, Circulation
     57: 968-975 (1978).
7.   H. Javid, H.M. Tufo, H. Jajafi, W.S. Dye, J.A. Hunter, and
     O.C. Julian, Neurological abnormalities following open heart
     surgery, J. Thorac. Cardiovasc. Surg. 58: 502-509 (1969).
8.   Y. Tamari, L. Aledort, E. Puskin, T.J. Degnan, N. Wagner,
     M.J. Kaplitt, and E.C. Pierce, Functional changes in plate-
     lets during extracorporeal circulation, Ann. Thorac. Surg.
     19: 639-647 (1975).
9.   R. McKenna, F. Bachmann, B. Whittaker, J.R. Gilson, and
     M. Weinberg, The haemostatic mechanism after open heart
     surgery II. Frequency of abnormal platelet functions during
     and after extracorporeal circulation, J. Thorac. Cardiovasc.
     Surg. 70: 298-308 (1975).
10. W.H. Lee, D. Krumbhaar, E.W. Fonkalsrud, O.A. Schjeide, and
     J.V. Maloney, Denaturation of plasma proteins as a cause of
     morbidity and death after intracardiac operations, Surgery
     50: 29-39 (1961).
11. J. Feijen, Thrombogenesis caused by blood-foreign surface
     interaction, in: "Artificial Organs," R.M. Kenedi, J.M.
     Courtney, J.D.S. Gaylor, and T. Gilchrist, ed, p.p. 235-247,
     University Park Press, Baltimore (1977).
12. G.N. Olinger, R.M. Becker, and L.I. Bonchek, Non cardiogenic
     pulmonary oedema and peripheral vascular collapse following
     cardiopulmonary bypass. Rare protamine reaction?, Ann.
     Thorac. Surg. 29: 20-25 (1980).
13. D.J. Parker, S.W. Cantrell, R.B. Karp, R.M. Stroud, and S.B.
     Digerness, Changes in serum complement and immunoglobulins
     following cardiopulmonary bypass, Surgery 71: 824-827 (1972).
14. P. Hairston, J.P. Manos, C.D. Graber, and W.H. Lee, Depression
     of immunologic surveillance by pump oxygenation perfusion,
     Surg. Res. 9: 587-593 (1969).
15. T.E. Hugli, Complement anaphylatoxins as plasma mediators,
     spasmogens and chemotaxins, in: "Current topics in molecular
     immunology," R.N. Reisfield, and W.J. Mandy, ed, p.p. 255-279
     Plenum Press, New York (1979).
16. D.E. Hammerschmidt, P.R. Craddock, J. McCullough, R.S.Kronenberg
     A.P. Dalmasso, and H.S. Jacob, Complement activation and
     pulmonary leukostasis during nylon fibre filtration leuka-
     phorasis, Blood 51: 721-30 (1970).
17. P.R. Craddock, J. Fehr, K.L. Brigham, R.S. Kronenberg, and H.S.
     Jacob, Complement and leukocyte-mediated pulmonary dys-
     function in haemodialysis, N. Engl. J. Med. 296: 769-774
     (1974).

18   D.E. Hammerschmidt, D.F. Stroncek, T.K. Bowers, C.J. Lanni-
      Keefe, D.M. Kurth, A. Ozalins, D.M. Nicoloff, R.C. Lillehei,
      P.R. Craddock, and H.S. Jacob, Complement activation and
      neutropenia occurring during cardiopulmonary bypass, J.
      Thorac. Cardiovasc. Surg. 81: 370-377 (1981).

19.   S.W. Fountain, B.S. Martin, C.E. Munsclow, and J.D. Cooper,
      Pulmonary leukostasis and its relationship to pulmonary
      dysfunction in sheep and rabbits, Circ. Res. 46: 175-180
      (1980)

20.   T.E. Hugli, and D.E. Chenoweth, Biologically active peptides
      of complement. Techniques and significance of C_3a and C_5a
      measurements, in: "Future perspectives in clinical laboratory
      immunoassays," R.M. Nakamura, W.R. Dito and E.S. Tucker, ed,
      p.p. 443-460 Alan R. Liss, New York (1981).

21.   D.E. Chenoweth, S.W. Cooper, T.E. Hugli, R.W. Stewart, E.H.
      Blackstone, and J.W. Kirklin, Complement activation during
      cardiopulmonary bypass. Evidence for generation of C_3a and
      C_5a anaphylatoxins, N. Engl. J. Med. 304: 497-503

22.   R. Rent, N. Ertel, R. Eisenstein, and H. Gewurz, Complement
      activation by interactions of polyanion and polycations I.
      Heparin-protamine induced consumption of complement, J.
      Immunol. 114: 120-124 (1975).

23.   J.K. Kirklin, S. Westaby, E.H. Blackstone, J.W. Kirklin, D.E.
      Chenoweth, and A.D. Pacifico, Complement and the damaging
      effects of cardiopulmonary bypass, J. Thorac. Cardiovasc.
      Surg. 86: 845-857 (1983).

24.   R.L. Replogle, A.B. Gazzaniga, and R.E. Gross, Use of corti-
      costeroids during cardiopulmonary bypass. Possible lysosomal
      stabilisation, Circulation 33 and 34 (suppl. I): 86-92 (1966)

25.   T. Sacks, C.F. Moldow, P.R. Craddock, T.K. Bowers, and H.S.
      Jacob, Oxygen radicals mediate endothelial cell damage by
      complement-stimulated granulocytes, J. Clin. Invest. 61:
      1161-1167 (1978).

ASSESSMENT OF PLATELET INHIBITORY THERAPY FOR ARTERIAL PROSTHESES

IN AN IN VIVO CANINE MODEL

Ian Lane, Joseph Irwin, Keith Poskitt and Charles McCollum

Department of Surgery
Charing Cross Hospital
Fulham Palace Road, London W6 8RF

Platelet inhibitory therapy has applications in vascular surgery and in patients who have coronary and cerebral vascular disease[1]. Screening of new products to assess their ethicacy involves measurement of platelet aggregation responses to standard aggregants[2] the use of artificial circuits[3] and clinical trials, which are time consuming and expensive. The frequent side-effects found with the standard platelet inhibitory therapy of aspirin combined with dipyridamole[4] has led for a search for new compounds. We have developed an in vivo canine model which is used to identify effective compounds on the basis of the rate of platelet accumulation and subsequent thrombosis on artificial arterial prostheses. We describe this model and its use to evaluate a new reversible cyclo-oxygenase inhibitor, indobufen (Farmitalia Carlo Erba).

METHODS

Twenty four greyhounds (weight 20-25kg) were allocated to receive indobufen 200mg, aspirin (ASA) 150mg plus dipyridamole (DPM) 50mg or placebo 12 hourly. Two days following the start of therapy, a 6 cm length of superficial femoral artery of one thigh was replaced by an equal length of 6mm diameter polytetrafluorethylene (Goretex) arterial graft (Fig 1). Seven days after surgery autologous platelets, labelled with 75 μ Ci of 111-Indium oxine, were re-injected. Over the subsequent 5 days the change in radioactivity over the graft was assessed using a highly collimated sodium iodide crystal probe and rate meter. Simultaneously, radioactivity of 5cc of blood was measured in a well counter. The daily increase in the ratio of radioactivity of graft over blood, termed thrombogenicity index (TI), was calculated[5]. Platelet survival was computed using a linear

Fig 1   The superficial femoral artery of the thigh has been
         replaced by a 6cm length of polytetrafluorethylene prosthesis

survival model[6]. Daily graft patency was determined, if necessary
using an ultrasound probe, and at 28 days the grafts were excised.
The grafts were sectioned at the mid-point and 5mm from each
anastomosis and under the light microscope the quantity of intraluminal thrombus expressed as a percentage of the total cross-
sectional area. For each graft the mean of the 3 individual
readings was calculated. Results were compared statistically using
the Mann-Whitney U-test.

RESULTS

     All dogs tolerated the medication and surgery. Mean
thrombogenicity index for the placebo treated group was 0.33 ± 0.07
(se mean). Thrombogenicity index was reduced to 0.14 ± 0.01 by
indobufen and 0.19 ± 0.04 by ASA + DPM (Fig 2). In the case of
indobufen the reduction in TI when compared to placebo was
significant ($p<0.05$). Two of the grafts in the placebo group
thrombosed before thrombogenicity index measurements could be
completed and are thus not included in the statistical analysis.
Platelet survival as determined by linear regression was 155 hours
(placebo), 133 hours (indobufen) and 146 hours (ASA + DPM). There
was no statistical difference in the figures.

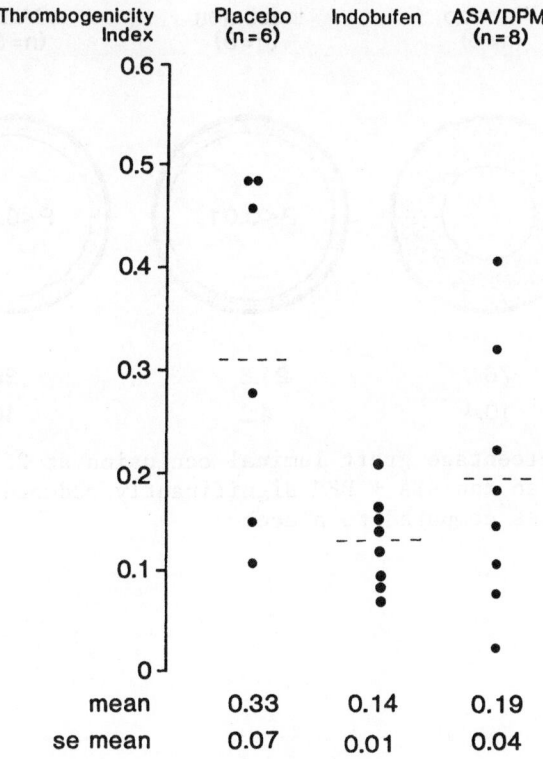

	Placebo (n=6)	Indobufen	ASA/DPM (n=8)
mean	0.33	0.14	0.19
se mean	0.07	0.01	0.04

Fig 2   Scattergram showing the thrombogenicity index measurements
for each dog within the 3 groups.

The graft luminal thrombus thickness at 28 days was
significantly reduced (p<0.02) by both platelet inhibitory therapies
when compared to placebo (Fig 3).   Three of the 8 placebo grafts
were patent at 28 days compared to 8 in the dogs treated with
indobufen and 7 in those treated with aspirin and dipyridamole.

DISCUSSION

The use of an in vivo animal model to assess platelet inhibitory
drugs examines the natural incorporation of red blood cells and
fibrin into the developing thrombus, in addition to platelets.   This
is not achieved by in vitro aggregation studies and artificial[3]
circuits where heparin is required as an anti-coagulant and may
have a platelet inhibitory effect itself.   The first month
following implantation of an arterial prosthesis is associated with

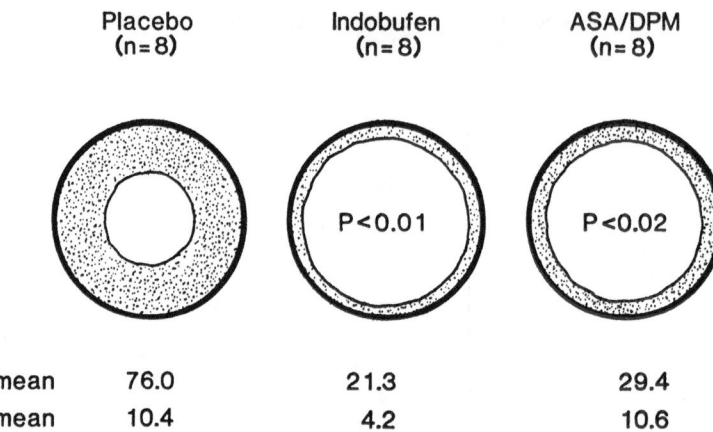

Placebo (n=8)	Indobufen (n=8)	ASA/DPM (n=8)

	Placebo	Indobufen	ASA/DPM
mean	76.0	21.3	29.4
se mean	10.4	4.2	10.6

Fig 3   Mean percentage graft luminal occlusion at 28 days. Both indobufen and ASA + DPM significantly reduced thrombus thickness compared to placebo.

an intense intraluminal thrombotic response which would be influenced by even minor platelet inhibitory effects. Both platelet accumulation (thrombogenicity index) and thrombus development are assessed independently in this model.

Indobufen was at least as effective as the combination of aspirin and dipyridamole in reducing platelet deposition in these Goretex grafts in dogs. It also significantly reduced intraluminal thrombus thick   at 28 days, although in the number of grafts studied the increase in graft patency did not achieve statistical significance. Surprisingly, platelet survival was slightly lowered by both therapies and this is in contrast to previous studies[7].

Further evaluation of indobufen is required in patients and if it proves to have few side effects it may have wide applications in the inhibition of thrombotic and embolic events in man. The canine arterial graft model is a reliable means of screening for platelet inhibitory activity as has been shown by the reduction in luminal thrombus thickness by the standard combination of aspirin and dipyridamole. Results are achieved within 28 days of the start of a study. This model will continue to be used for the evaluation of platelet inhibitory drugs.

REFERENCES

1. E.C. Tsu, Antiplatelet drugs in arterial thrombosis: A review,
    Am. J. Hosp. Pharm. 35: 1507, (1978).
2. J.F. Mustard and M.A. Packham, Factors influencing platelet
    function, adhesion release and aggregation, Pharmacological
    Rev. 22: 97, (1970).
3. C.N. McCollum, M.J. Crow, S.M. Rajah and R.C. Kester, Anti-
    thrombotic therapy for vascular prostheses: an experimental
    model testing platelet inhibitory drugs, Surgery 87: 668
    (1980).
4. PARIS Reserach Group, Persantine and aspirin in coronory heart
    disease, Circulation 62, 668, (1980).
5. M. Goldman, H.C. Norcott, R.J. Hawker, C. Hall, Z. Drolc and
    C.N. McCollum, Femoropopliteal bypass grafts - an isotope
    technique allowing in vivo comparison of thrombogenicity,
    Br. J. Surg. 69: 380, (1982).
6. E.A. Murphy and D.R. Bolling, Platelet Survival, in: "Platelet
    Function Testing", H.J. Day, H. Holmsen, M.B. Zucker eds.,
    US Department of Health Education and Welfare, (1978).
7. J.L. Ritchie and L.A. Harker, Platelet and fibrinogen survival
    in coronary atherosclerosis, Am.J. Cardiol. 39: 595 (1977).

EVOLUTION OF THE SUBENDOTHELIAL GROWTH ON

HEPARINISED ALDEHYDE CROSSLINKED ARTERIES

Madeleine Moczar, Philippe David*, and Daniel Loisance*

Laboratoire de Biochimie du Tissu Conjonctif, CNRS Gr40
*Centre de Recherches Chirurgicales, CNRS UA 591
Faculté de Médecine, Paris XII, Créteil 94000, France

INTRODUCTION

The macromolecular constituents of prostheses of biological origin are recognised by the hydrolytic enzymes of the host and degraded in vivo. The host cells respond to the implantation and biodegradation of vascular substitutes by the synthesis of intra-luminal and/or periprosthetic tissues. A neointimal proliferation progressing to an intraluminal occlusion may be one of the reasons of the long term postoperatory failures. Prosthetic aortic allo-grafts in rats were replaced by an endothelialized elastic tissue, patent to blood[1,2]. The biosynthetic pattern of proteins in the elastin containing scar was similar to that of proteins in the intima-media of host aorta[1].

The aim of the present work was to obtain informations on the extracellular matrix synthetized in response to heterologous heparinised bioprostheses. The prosthetic replacement of blood vessels with internal diameter inferior to 3 mm is still an un-resolved problem, thus our biochemical experiments were focused on the evolution of the tissular growth on small caliber pros-theses from human placenta arteries. The conduits were implanted in rat aortas and the macromolecular matrix synthetized de novo was investigated in terms of the biosynthetic labelling of gly-cosaminoglycans at three months and one year following aortic replacement.

MATERIAL AND METHODS

The flow sheet of the experiments is outlined in Fig. 1

137

Prostheses

Human placenta arteries of about 1 mm diameter removed in 6 hrs following normal deliveries were washed with sterile 0.9 % NaCl, supplemented with antibiotics. The samples were treated with heparin and crosslinked with glutaraldehyde. Conduits of about 5 mm length were implanted into infrarenal aorta of Wistar rats by microsurgical techniques[3]. The conduits were assayed for hydrolysis with pepsin and bacterial collagenase[1] and for the leaking out of heparin at increasing ionic strengths.

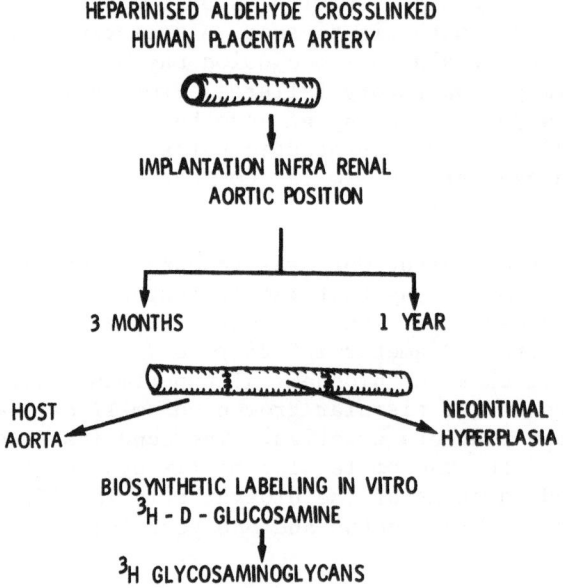

HEPARINISED ALDEHYDE CROSSLINKED
HUMAN PLACENTA ARTERY

IMPLANTATION INFRA RENAL
AORTIC POSITION

3 MONTHS   1 YEAR

HOST   NEOINTIMAL
AORTA   HYPERPLASIA

BIOSYNTHETIC LABELLING IN VITRO
^{3}H - D - GLUCOSAMINE

^{3}H GLYCOSAMINOGLYCANS

Fig. 1. Experimental scheme of the long term follow up of prostheses from human placenta arteries. Aortic substitutions in rats (n = 78)

Biosynthetic labelling

At three months and one year following aortic replacement the blood permeable conduits with the subendothelial growths

and the host aortas were excised. The hyperplasias and host
aortas from six rats were pooled by two and incubated with
50μCi D- [1 ³H] glucosamine (17 Ci/mMol) in Eagle's minimum essential medium for 4 hrs with continous shaking. The ³H glucosamine
labelled samples were hydrolysed with collagenase and pronase and
the ³H glycosaminoglycans were identified by hydrolysis with chondroitin AC lyase, chondroitin ABC lyase and deaminative cleavage
with nitrous acid[4].

RESULTS AND DISCUSSION

     The collagen in the fresh, untreated placenta arteries was
quantitatively solubilised with pepsin and degraded with bacterial
collagenase. The effect of these enzymes on collagen in the glutaraldehyde crosslinked arteries was not detectable in vitro.
About 3 % of the amount of heparin associated with the prostheses
could be extracted with 0.15 M NaCl at 37° C. The removal of the
residual heparin with 1.5-2 M NaCl solution in vitro, indicated
that the heparin was bound to the conduits by ionic forces.

     The patency was 60 % at three months and one year after the
implantation of the conduits. Sparse deposition of fibrin and
individual platelets were observed on the prosthetic surface by
scanning electron microscopy in the first 40 days following aortic
substitution. At 3 months subendothelial hyperplasia grown from
the both ends of the anastomoses[3] has covered the whole prosthetic
luminal surface in 70 % of the blood permeable conduits.

     Biochemical studies were performed to characterize the newly
synthetised tissue and the behaviour host aortic cells in their
environment modified by the prosthetic implant. Aortic smooth
muscle cells and endothelial cells exhibit a characteristic biosynthetic pattern of glycosaminoglycans[5]. The metabolic labelling
of glycosaminoglycans, components of the pericellular and extracellular compartments provides an experimental approach to assess
the evolution of the neointimal growth and the host aorta. This
biochemical technique is based on the biosynthetic activity of
cells, thus the acellular prosthetic remnants adhering to the
neointimal hyperplasia do not interfere in the analytical assays.
The specific radioactivity of glycosaminoglycans were
3000 ± 500 ³H cpm/μg uronic acid and 400 ± 100 ³H cpm/μg uronic
acid in the subendothelial growth at three months and one year
respectively. At one year, the incorporation of ³H glucosamine
into glycosaminoglycans on uronic acid basis was similar in the
neointima and in the host intima-media. The results reflect,
that the biosynthetic activity of cells involved in the repair of
aortic injury decreased to the level found in the host aorta and
in the control,unoperated aortas at one year following prosthetic

replacement. The incorporation patterns of ^{3}H glucosamine into glycosaminoglycans in the subendothelial hyperplasia and host aorta at three months and one year following aortic substitution are given in Fig. 2.

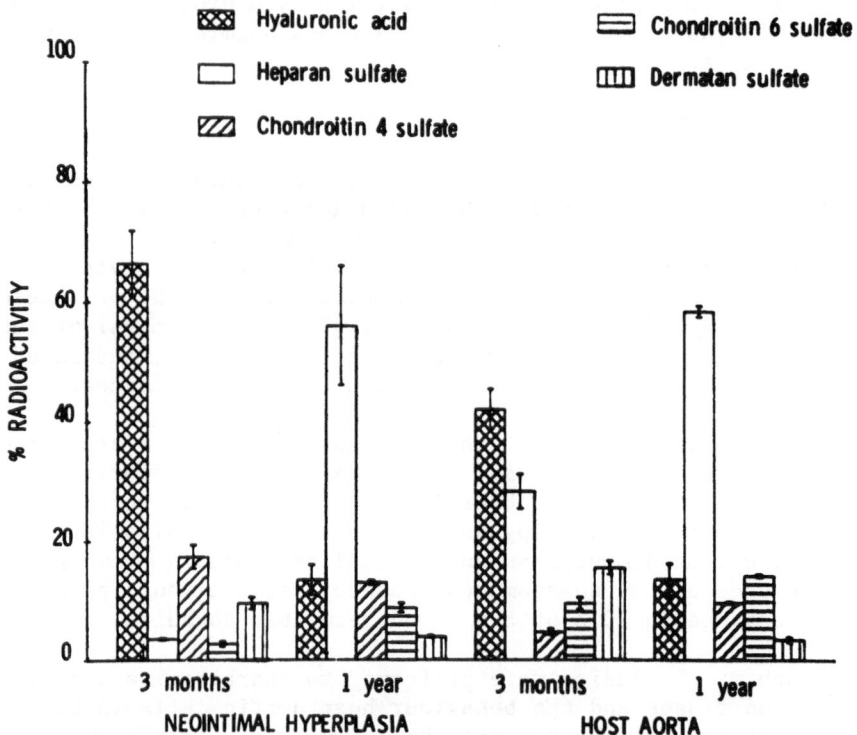

Fig. 2. Biosynthetic labelling pattern of glycosaminoglycans in the subendothelial growth and in the intima-media of rat host aorta at three months and one year following aortic substitution with glutaraldehyde crosslinked human placenta arteries. Results are expressed as a percentage of total radioactivity incorporated into glycosaminoglycans.

The distribution of ^{3}H glycosaminoglycans was similar in the unoperated control aortas and in the host intima-medias. These data let us assume, that the biosynthetic activity of distal host aortic cells was not modified by the implantation of the conduits. At three months about 60 % and 42 % of the ^{3}H glycosaminoglycans were sulfated in the host aorta and the neointima respectively. In the neointima, the highest ^{3}H glycosaminoglycan label was identified as hyaluronic acid at 3 months. One explanation of the relative high hyaluronic acid synthesis could be the presence of migrating cells[6] in the neointimal hyperplasia at the early phase of the evolution. From three months to one year, the ^{3}H label of hyaluronic acid decreased about fivefold in the neointima concomitantly to the decrease of the ratio of ^{3}H chondroitin 4 sulfate to ^{3}H chondroitin 6 sulfate. A parallelism may be found between the increased label of chondroitin 6 sulfate in the neointima at one year and the increased chondroitin 6 sulfate content of connective tissues with age[7]. The results on the incorporation of ^{3}H glucosamine into glycosaminoglycans indicate that the composition and metabolism of the extracellular matrix in the neointimal hyperplasia and host aorta undergo changes with progressing time in vivo.

The time dependent modifications are in the favour of a maturation process in the neointima paralleled with the ageing of the host aorta. The results in experimental surgery[3] and the biochemical findings let us assume, that the heparinised aldehyde crosslinked human placenta arteries were well tolerated as heterologous vascular substitutes. The hyaluronic acid rich neointimal proliferation replacing the prosthetic remnants showed a tendency to evoluate to a "media like" tissue in the blood permeable conduits.

REFERENCES

1. M. Moczar, J.P. Bessou and D. Loisance. Healing of biodegradable vascular prosthesis. Incorporation of ^{3}H valine into proteins in the subendothelial scar and host intima-media of rat aorta. Connective Tissue Research 12:33 (1983)
2. D. Loisance, M. Moczar, J. Leandri, J.P. Bessou and Ph. David. A new microarterial biograft. Trans. Am. Soc. Artif. Intern. Organs 27:401 (1981)
3. D. Loisance, Ph. David, J. Leandri, M. Moczar. Etude experimentale d'une nouvelle prothèse micro-artérielle. J. Chirurgie 121:355 (1984)
4. M. Moczar, Ph. David and D. Loisance. Vascular substitute from human placenta arteries. Glycosaminoglycan and elastin synthesis in neointimal hyperplasia. Life Support Systems 2:201 (1984)
5. G. Gamse, H.G. Fromme and H. Kresse. Metabolism of sulfated glycosaminoglycans in cultured endothelial cells and smooth muscle cells from bovine aorta. Biochim. Biophys. Acta 544:514 (1978)
6. B.P. Toole. Glycosaminoglycans in morphogenesis. In "Cell Biology of the Extracellular Matrix" E. D. Hay, ed. Plenum Press, New-York (1982)

7. L. Robert and M. Moczar. Age related changes of proteoglycans
    and glycosaminoglycans. In "Glycosaminoglycans and Proteo-
    glycans in Physiological and Pathological Process of Body
    Systems". R.S. Varma, R. Varma and Pa. Warren, eds Karger,
    Basel (1982)

THE INFLUENCE OF THROMBOXANE RECEPTOR BLOCKADE ON PLATELET UPTAKE

IN DACRON GRAFTS IN MAN

Ian Lane, Marion Sinclair, Keith Poskitt, and
Charles McCollum

Department of Surgery
Charing Cross Hospital
Fulham Palace Road, London W6 8RF

Human prosthetic vascular grafts do not sustain a growth of endothelium on the luminal surface and remain thrombogenic indefinitely[1]. Whilst large diameter aortic grafts develop a thin layer of platelet thrombus as pseudointima, smaller prostheses such as those in the femoro-popliteal position have a thrombosis and occlusion rate approaching 60% at 1 year after implantation[2]. Platelet inhibitory therapy is established in the prevention of thrombosis but the combination of aspirin plus dipyridamole produces frequent gastro-intestinal side effects. This has been shown in the recent Persantin Aspirin Re-infarction Study[3] where 25% of patients had to discontinue aspirin therapy. Thromboxane A_2 (TXA_2) which is a product of aracidonic acid metabolism, is a powerful stimulator of platelet aggregation[4]. It is produced by enzymes including cyclo-oxygenase and thromboxane synthetase in platelets and the natural biological opponent of TXA_2 is prostocyclin (PGI_2).

We have investigated the effect of a novel competitive thromboxane A_2 antagonist, AH23848B (Glaxo Research Limited), on platelet accumulation in mature Dacron aorto-bifemoral grafts.

PATIENTS AND METHODS

Thirty patients with Dacron aorto-bifemoral grafts implanted more than 1 year previously were randomly allocated to receive AH23848B 70mg, aspirin (ASA) 300mg plus dipyridamole (DPM) 75mg, or placebo 8 hourly for 10 days. No patient had been treated by platelet inhibitory therapy for 1 month prior to entry into the study. The ratio of the concentration a thromboxane mimetic

(U46619) needed to achieve 50% platelet aggregation in whole blood during treatment to the concentration required prior to therapy (CR50) was determined[5]. Five days following the initiation of therapy autologous platelets labelled with 150 µ Ci of 111-Indium oxine were injected. Over the next 5 days the accumulation of radioactive platelets on the graft was measured by placing a highly collimated probe containing a sodium iodine crystal over the graft and gamma camera imaging. At the same time the radioactivity of 5cc of whole blood was measured by a well crystal attached to a rate meter. The daily rise in the ratio of radioactivity over the graft to that of blood was expressed as the thrombogenicity index (TI)[6]. Platelet survival was determined by using a multiple hit model[7]. Results were compared statistically using the Mann-Whitney U-test.

RESULTS

All patients completed the study and no side effects were observed. There were 27 men and 3 women; mean age 64.8 years. Mean graft age was 5.4 ± 0.6 (se mean) years. All patients receiving the thromboxane antagonist showed a mean concentration ratio (CR50) to aggregant U46619 of 249 ± 99. This was significantly higher than the placebo value of 1.1 ± 0.2 (Fig 1). In the aspirin plus dipyridamole group there was only a small increase in concentration ratio compared to the placebo. It is known that aspirin does not interfere with the aggregation pathway activated by thromboxane mimetics.

Mean thrombogenicity index in the placebo group was 0.30 ± 0.06 (Fig 2). This was reduced to 0.16 ± 0.03 in the case of the thromboxane antagonist (p<0.05) and to 0.28 ± 0.06 for aspirin plus dipyridamole. Gamma images performed 4 days after the injection of the indium labelled platelets, in patients treated with placebo, showed an accumulation of radioactivity when compared to the images obtained immediately after injection (Fig 3). This accumulation was not so marked in the group treated with the thromboxane antagonist. There was no significant difference in platelet survival between the 3 groups.

DISCUSSION

The thromboxane A_2 antagonist AH23848B significantly reduced the rate of platelet accumulation on human Dacron grafts. This effect was not seen with the standard combination of aspirin plus dipyridamole. High doses of aspirin reduce prostacyclin production by the arterial endothelium[8]. Prostacyclin inhibits platelet aggregation and is the biological opponent of thromboxane A_2. However, there is no evidence that prostacyclin is secreted by the

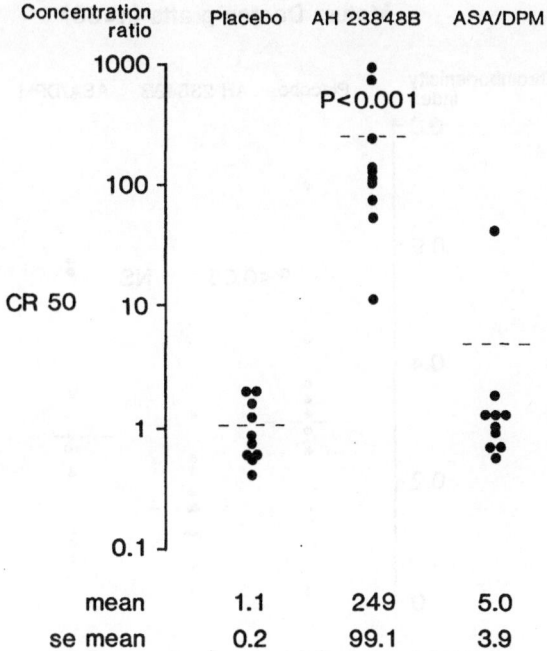

Figure 1.   Platelet aggregation to U46619 expressed as CR50
for the three groups. Mean CR50 for the patients
treated with AH23848B was significantly increased
when compared to placebo.

neointima of an arterial prostheses. Thus reduction of prostacyclin
synthesis may not be the reason of the high ethicacy of the thromboxane antagonists in this study. Dazoxiben, a selective thromboxane synthetase inhibitor, also appears to have the advantage of
directing thromboxane precursors towards prostacyclin synthesis.
This drug has shown encouraging results in clinical trials [9]
although it may also divert these thromboxane precursors towards
the pro-aggregatory protanoids such as PGG_2 and PGH_2. There were
no side effects in the 10 patients treated with the thromboxane
antagonist and if the trend is continued this drug represents one
of an important new group of compounds which may have wide application where platelet inhibitory therapy is required.

Mature Dacron grafts (n=30)

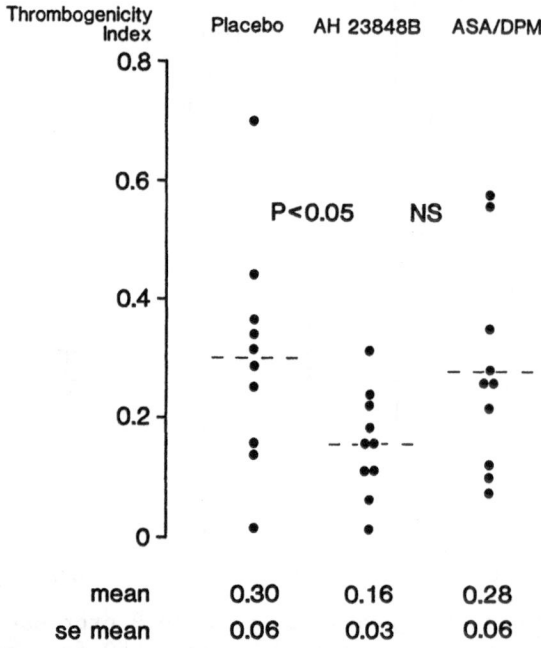

Fig 2   Thrombogenicity Index measurements plotted individually for
         each patient in the three groups. Mean TI for AH23848B
         was significantly lower (p<0.05) than placebo.

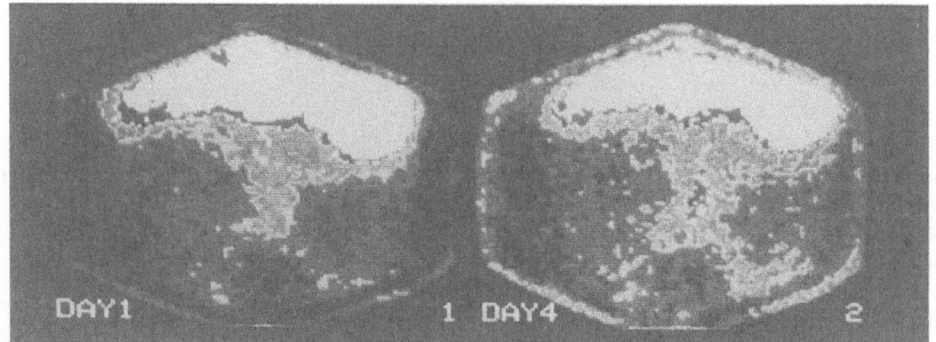

Figure 3.   Gamma camera images of the abdomen of a patient treated
             with placebo, obtained at Day 1 and Day 4 following
             injection of III-Indium labelled platelets. There is
             increased accumulation of radioactivity of the aorto-
             bifemoral graft on Day 4 when compared to Day 1.

REFERENCES

1.   M. Goldman, H.C. Norcott, R.J. Hawker, Z. Drolc and C.N.
      McCollum, Platelet accumulation on mature Dacron grafts
      in man, Br. J. Surg. 69 (Suppl.): 538, (1982).
2.   I.G. Kishon, D.W. Stoney, D.J. Tibbs and P.J. Morris, Expanded
      polytetrafluoroethylene grafts for severe lower limb
      ischaemia, Br.J.Surg. 68: 173,(1981).
3.   Paris Research group, Persantin and aspirin in coronary heart
      disease, Circulation, 62: 449, (1980).
4.   M. Hamberg, J. Svensson and B. Samuelson, Thromboxanes - a new
      group of biologically active compounds derived from prosta-
      glandin endoperoxides, Proc. Natl. Acad. Sci. USA. 72: 2994,
      (1975).
5.   P. Lumley and P.P.A. Humphrey, A method for quantitating plate-
      let aggregation and analysing drug-receptor interactions on
      platelets in whole blood in vitro, J. Pharmacol. Meth. 6: 153
      (1981).
6.   M. Goldman, H.C. Norcott, R.J. Hawker, C. Hall, Z. Drolc and
      C.N. McCollum, Femoro-popliteal bypass grafts - an isotope
      technique allowing in vivo comparison of thrombogenicity,
      Br. J. Surg. 69: 380 (1982).
7.   D.J. Doyle, C.N. Chesterman, J.F. Casle and F.J. Morgan,
      Platelet concentrations of platelet specific proteins correl-
      ated with platelet survival, Thromb. and Haemost. 42: 329
      (1979).
8.   E.F. Ellis, K.F. Wright, P.J. Jones, D.W. Richardson and C.K.
      Ellis, Effect of oral aspirin dose on platelet aggregation
      and vascular prostacyclin (PGI) synthesis in humans and
      rabbits, J. Cardiovas. Pharmac. 2: 387 (1980).
9.   M. Goldman, C. Hall, R.J. Hawker and C.N. McCollum, Dazoxiben
      examined for platelet inhibitory effect in an artificial
      circulation, Br. J. Pharmac. 15:615 (1983).

PLASMA FIBRONECTIN AND PLATELET DEFICIENCY FOLLOWING AORTIC CROSS-CLAMPING

J.T. Powell*, K.R. Poskitt, J.T.C. Irwin and C.N. McCollum

Departments of Surgery and Biochemistry
Charing Cross and Westminster Medical School
London, W6 8RF, U.K.

Fibronectin is an adhesive glycoprotein present in high concentration in plasma, >300 mg/l. Fibronectin has binding sites for collagen, heparin, cells, Factor XIII and other macromolecules.[1] Plasma fibronectin is thought to function as a non-specific opsonin to promote reticuloendothelial cell clearance of microaggregates. The depletion of this opsonin has been related to poor prognosis in critically ill surgical and burn patients.[2] Plasma fibronectin depletion has also been reported in the early postoperative period in patients undergoing uncomplicated gastro-intestinal surgery.[3] We have developed a model of surgical shock in pigs involving aortic-cross clamping and followed the plasma fibronectin and platelet counts.

EXPERIMENTAL

Fourteen pigs (30±4 kg) were subject to laparotomy under infusion anaesthesia with Brietal. Infrarenal aortic clamps were placed and the aorta held clamped for 1½ hours, the small bowel being exteriorized for 30 min. Following declamping a further period of 1 hour shock ensued before resuscitation with intravenous fluids. Venous blood was taken from a venous cannula on the day prior to operation, pre-operatively, after placement of the aortic clamp, after declamping, after resuscitation and 3 days postoperatively. Two control pigs were subject to anaesthesia without laparotomy. Fibronectin samples were collected in EDTA bottles supplemented with 3 mM benzamidine as a protease inhibitor. Plasma was stored at -20°C prior to analysis by ELISA for fibronectin;[4] porcine plasma

149

fibronectin purified on gelatin-Sepharose was used as a standard.
The presence of fibronectin complexes was assessed by molecular
sieve chromatography on Sepharose CL 4B (1 x 60 cm) developed with
phosphate buffered saline. The reactivity of fibronectin samples in
the ELISA assay was measured by serial dilution and the addition of
exogenous fibronectin. Statistical analysis was by paired Wilcoxon
rank test, all resuls expressed as mean ± SEM.

RESULTS

The change in platelet count and plasma fibronectin are shown
in Table 1 for the pigs subject to aortic clamping. The platelet
count fell steadily from the preoperative value of 487 ± 61 x 10^9/l
to 98 ± 13 x 10^9/l 3 days postoperatively. The fibronectin also
fell steadily on the day of operation but recovered to near normal
3 days postoperatively to 245 ± 45 mg/l. In the two control pigs
there was no sigificant change in either platelet count or plasma
fibronectin level.

On the day of operation the platelet count correlated significantly with the plasma fibronectin (r ≠ 0.57, p < 0.001). This
correlation was not maintained at 3 days post laparotomy since
platelet levels continued to fall but plasma fibronectin had recovered. Only 7 of the 14 shocked pigs survived at 3 days post
laparotomy. The severity of the fibronectin depletion related to
subsequent survival: following resuscitation the plasma fibronectin
fell to 24 ± 6 mg/l in the 7 pigs that died within 24 hours of
laparotomy compared to 96 ± 16 mg/l in the 7 survivors, p < 0.05.

Table 1

	Fibronectin and platelet count in pigs subject to aortic clamping			
	Pre-operative	Aorta clamped	Aorta unclamped	After resuscitation
Fibronectin (mg/l)	331±10	234±18*	133±20*	43±13*
Platelet count (x 10^{-9}/l)	487±61	363±55*	293±46*	259±54*

*p<0.01 compared to pre-operative count

The platelet count after resuscitation did not distinguish so clearly the two groups.

Following declamping of the aorta and resuscitation fibronectin was found to elute from Sepharose chromatography with an estimated M_r $\sim10^6$ (Figure 1). Preoperative plasma fibronectin eluted in a single peak, M_r 450,000. The addition of exogenous fibronectin

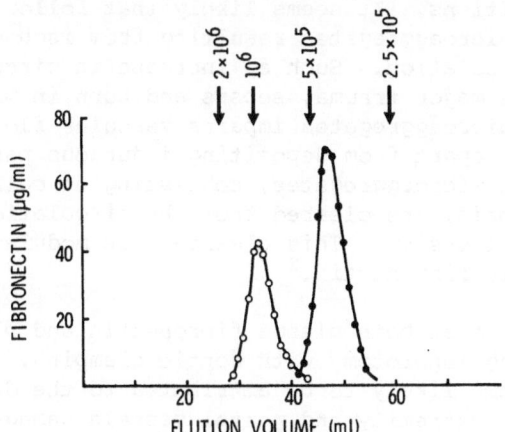

Fig. 1. Molecular sieve chromatography of fibronectin on
Sepharose CL 4B.
The column (1 x 65 cm) was developed with phosphate
buffered saline and calibrated with marker proteins:
-●-, preoperative plasma, -O-, plasma after aortic
declamping assayed by ELISA for fibronectin.

(50-200 µg/ml) to plasma obtained after aortic declamping resulted in the conversion of all fibronectin to the high molecular weight species (M_r $\sim10^6$). The reactivity of fibronectin in the ELISA assay was very similar for both 450 K and 1000 K forms.

These results suggest that firstly we are observing a true
depletion of plasma fibronectin in the shocked pigs, secondly that
all the fibronectin is consumed in high molecular weight complexes,
there is no evidence for proteolytic fragments, and thirdly that
excess of the species that complexes to fibronectin remains in the
plasma after aortic declamping. Preliminary experiments suggest
that collagen is present in the high molecular weight complexes
found on aortic declamping.

DISCUSSION

Extensive operations with aortic clamping are routine in
cardiovascular surgery. Both clamping and the shock of surgery lead
to lower limb ischaemia. We have developed an animal model that
mimics these conditions. It seems likely that following the release
of aortic clamps microaggregates resulting from ischaemic injury
will enter the circulation. Such an increase in circulating microaggregates follows major trauma, sepsis and burn in man. The persistence of such microaggregates impairs vascular flow and impedes
lymphatic drainage apart from depositing injurious particles on the
endothelium. Such microaggregates, consisting of cellular and connective tissue debris, are cleared from the circulation by the
reticuloendothelial system. This clearance is modulated by humoral
factors, especially fibronectin.[2]

We have shown that both plasma fibronectin and platelet counts
fall rapidly during laparotomy with aortic clamping. The fall in
platelet count seems likely to be attributed to the deposition of
platelets on lower extremity and portal vessels damaged by ischaemia
and to the formation and clearance of platelet aggregates. The
dramatic fall in plasma fibronectin, together with its change in
size, suggests that fibronectin is being consumed in opsonic complexes. From preliminary evidence it appears that these fibronectin
complexes may contain collagen, probably deriving from damaged microvasculature. The extent of fibronectin depletion would therefore
be expected to correlate with the amount of microaggregates released
from damaged tissue. Fibronectin depletion was also predictive of
mortality, the levels after resuscitation in the survivors being
three times those in the pigs that died.

Further research is needed to clarify the mechanism of fibronectin consumption, its relationship to microaggregate formation
and its influence on subsequent survival.

These reductions in platelet and fibronectin levels following
aortic clamping may have pathological significance in the generation
and clearance of microemboli and microaggregates. As fibronectin
promotes cell adhesion[1] and platelets deposit on prosthetic grafts[5]
both these factors are likely to be relevant to the cellular

colonization and patency of vascular prostheses. Studies in patients receiving aortic grafts are in progress to study the platelet and fibronectin profiles.

SUMMARY

Aortic surgery in pigs produces shock. Both the levels of circulating platelets and plasma fibronectin are profoundly depleted following aortic cross clamping. The plasma fibronectin is consumed in opsonic complexes and the extent of its depletion is related to mortality.

REFERENCES

1. R.O. Hynes, and K.M. Yamada, Fibronectins: Multifunctional modular glycoproteins, J. Cell Biol. 95: 369 (1982).
2. T.M. Saba, and E. Jaffe, Plasma fibronectin (opsonic glyco-protein): its synthesis by vascular endothelial cells and role in cardiopulmonary integrity after trauma as related to reticuloendothelial function, Am. J. Med. 68: 577 (1980).
3. S.J.D. Chadwick, J.F. Mowbray, and M.A.F. Dudley, Plasma fibronectin and complement in surgical patients, Br. J. Surg. 71: 718 (1984).
4. E. Ruosliolati, E.G. Hayman, M. Pierschbacher, and E. Engvall, Fibronectin, Methods Enzymol. 82: 803 (1982).
5. M. Goldman, C. Hall, J. Dykes, R.J. Hawker, and C.N. McCollum, Does [111]Indium-platelet deposition predict patency in prosthetic arterial grafts? Br. J. Surg. 70: 635 (1983).

SECTION 4

CELL SURFACE INTERACTIONS AND TUMOUR METASTASES

CELL SURFACE ALTERATIONS AND THE METASTATIC BEHAVIOR OF TUMOR CELLS

George Poste

Smith Kline and French Laboratories
Philadelphia, Pennsylvania 19101
U.S.A.

INTRODUCTION

The cell biology of cancer metastasis has been studied in considerable detail over the last ten years yielding many important new insights into the pathogenesis of this important and life-threatening disease which have been well reviewed in recent years [1-3]. In contrast, efforts at understanding the underlying biochemistry and molecular biology of this process have been limited, and largely unsuccessful. A much greater understanding of the molecular mechanisms responsible for the expression of the metastatic phenotype will be required if we are to develop novel and productive approaches to the therapy of metastatic disease by any approach other than random and semi-empirical screening, which hitherto has made little impact in developing anticancer drugs against the more common solid malignancies of man. In the absence of any obvious biochemical target to guide the direction of such an endeavor it is becoming increasingly clear that the process of identifying molecular properties that correlate either qualitatively or quantitatively with metastatic potential will be a lengthy and challenging task and with no guarantee that any correlation, once identified, will be causal or will offer a suitably "unique" and pharmacologically exploitable target for the design of new therapeutic agents directed against it.

The probability that alterations in the surface properties of cancer cells contribute to their aberrant behavior has long attracted the interests of cancer researchers. This attention has been productive to the extent that the modern literature contains an interminable litany of differences in cell surface properties that have been detected at one time or another (with variable consistency) in human and animal tumor cell populations of diverse tissue origins,

ranging from rodent cell lines, many of which have been passaged
in vitro in different laboratories for years or even decades, to
human tumor cells isolated from newly excised tumors. Unfortunately,
the only safe conclusion that can be drawn from these studies is that
we cannot yet state with certainty how any specific cell surface
change contributes or is even relevant to complex, multifactorial
behavioral traits such as tumorigenicity, invasiveness or the ability
to metastasize. Indeed, few of the enormous number of publications
on the neoplastic cell surface pertain to metastatic tumor cells.
This is surprising, or perhaps disturbing, in view of the fact that
metastatic disease, not the primary tumor, is the major challenge in
clinical oncology. Even fewer studies have been undertaken in which
analyses of cell surface properties, conducted almost exclusively
within cultured cell populations in vitro, have been accompanied by
parallel in vivo studies of the behavior of the same cells. Instead,
most investigators have been content to cite historical reports, or
studies by others, as evidence of the in vivo behavior of the cells
in question. This assumes that the cells in their own laboratory
will not differ to any significant degree. As discussed in detail
below, there is increasing evidence that this assumption is unwarr-
ented and there is a danger that the value of the application of
sophisticated molecular biological techniques to analyze tumor cells
in vitro will be undermined by failure to assay the in vivo behavior
of the cell preparations being studied.

There is compelling circumstantial evidence to implicate changes
in the structural and functional organization of the cell surface in
the etiology of the altered behavior of malignant cells. Within the
primary tumor, alterations in the surface properties of tumor cells
are presumed to contribute to their escape from many of the regulatory controls to which normal cells are subject. The proliferation
of tumor cells is no longer effectively regulated by cell-cell contact
interactions and surface changes are resonably viewed as contributing
to the altered responsiveness of tumor cells to various autocrine,
paracrine and endocrine signals.

Alterations in plasma membrane transport processes are common
in tumor cells. Such changes may facilitate improved nutrient
utilization by tumor cells, thereby conferring an adaptive growth
advantage on these cells under conditions in which nutrient supply
is limiting. Changes in membrane transport functions can also
render tumor cells resistant to diverse classes of chemotherapeutic
agents, either by frustrating drug uptake or by promoting active
drug efflux. The abberant regulation of positional control in
invasive tumor cells that allows them to infiltrate surrounding
host tissues probably involves lesions in surface properties that
alter cellular cognitive responses to homologous and heterologous
cells. Similarly, progression of the invasion process, culminating
in the entry of malignant tumor cells into lymphatics and blood
vessels, involves a complex set of events in which changes in

surface properties almost certainly contribute to the defective control of cell locomotion that allows tumor cells to infiltrate normal
tissues and to destroy host tissues via the presumed release of tissue
degradative enzymes. In malignant tumors, these processes are augmented by the metastatic cascade in which the surface properties of
tumor cells are believed to play a major role in the dissemination
of tumor cells to distant organs, including, in certain tumors, the
apparent selective, non-random localization of tumor cells in particular "target" organs. Finally, at all stages in tumor progression,
the surface properties of neoplastic cells affect their susceptibility
to recognition and destruction by host defence mechanisms.

Surface alterations in tumor cells are also viewed logically as
potential targets for antineoplastic agents. Unfortunately, with
the exception of determinants that are expressed by tumor cells in a
temporally inappropriate fashion and are absent from non-neoplastic
cells of the same lineage, the majority of surface changes detected
in tumor cells to date are of a quantitative rather than qualitative
nature, thus limiting their value as targets for achieving selective
therapeutic destruction of neoplastic cells. However, with the rapid
progress in developing antibodies to human tumor cells, there are an
increasing number of reports of epitopes expression by tumor cells that
do not appear to be present on normal cells of the same tissue and
these may be of considerable value in cancer diagnosis and therapy.
These observations have led to proposals to use the shedding of such
determinants as diagnostic markers and/or their exploitation in
engineering antibody-mediated "targeting" of imaging and therapeutic
agents to tumor cells. Although these approaches are in their
infancy, and have perhaps generated more publicity than their clinical
success to date warrants, future refinements in these approaches will
almost certainly generate worthwhile improvements in the diagnosis,
staging and/or therapy of malignant disease.

The diversity of the surface changes detected in cancer cells,
their potential functional importance and their value as possible
targets for therapeutic assault have been discussed at length in
several recent reviews [3-11], and a further survey would be redundant.
This article will focus instead on the type of experimental approaches
needed to translate our current descriptive perspectives on cell
changes in cancer into a mechanistic framework in which specific
surface changes can be shown to have a causal correlation with particular aspects of tumor cell behavior. The long term goal of this
exercise is to define the particular subsets of tumor cell properties that are responsible for clinically important aspects of tumor
cell behavior such as metastasis and resistance to killing by therapeutic agents or host defences. With this knowledge it might then be
possible to embark on the rational design of novel therapeutic agents
directed against tumor cells that display these traits.

TUMOR MODELS

The use of tumor models has been of overwhelming importance in
research on the pathogenesis of neoplastic disease and the search
for new antitumor agents. This will not change in the forseeable
future. It is essential, however, that the criteria used in choosing
a tumor model be re-examined at regular intervals to determine if
the model still fulfills the original intention and whether it should
be discarded or refined to accommodate new advances in cancer biology
that are occurring at a rapid pace as a result of dramatic progress
in cell biology, immunology and molecular genetics.

Animal tumor models exhibit many important differences from
human neoplasms. These have been documented in numerous publications
3, 12-15. Prominent among these are: differences in body size and
lifespan; species variation in the anatomy and physiology of major
organ systems; variation in the contribution of different elements
of host immune and non-immune reactions in combating neoplasia;
and significant species variation in the absorption, distribution,
metabolism and excretion of therapeutic agents.

The extensive reliance on transplantable established tumor cell
lines, usually of rodent origin, as models for human tumors represents
a major source of potential irrelevancies and artefacts. Most human
solid malignancies grow relatively slowly and contain cells that
retain many of the properties of the cell type from which they
originated. In contrast, many of the more widely used tumor cells
represent established lines of human and animal origin that typically
contain rapidly growing, anaplastic cells that have little or no
phenotypic relationship even to the cell type from which they arose
originally, yet alone tumors originating in entirely different cell
types. There is an urgent need to pay greater attention to the
biological uniqueness of neoplasms arising in different cell line-
ages 15. The increasing attention given by many laboratories to
the development of model systems using freshly isolated human tumor
cells is particularly relevant and represents a major initiative in
increasing the technical sophistication of current experimental
approaches.

Irrespective of whether human or animal tumor cells are used,
we must begin to give more attention to the nature and to the extent
of phenotypic change that results from the selection and serial
propagation of tumor cells in vitro or in vivo, and how far such
changes render the cells unsuitable, and thus irrelevant, for det-
ecting the cellular alterations responsible for tumorigenicity and
metastasis and for evaluating cellular responses to antitumor
agents 2, 15.

No attempt will be made in this article to debate the merits
of one tumor model versus another. Each different application of

metastatic tumor models will almost certainly impose unique experimental requirements that must be fulfilled by the model selected. The strengths and weaknesses of the transplantable animal tumors that are commonly used in metastasis research have been discussed at length elsewhere [2,9,11,15]. Each may have merits for the purpose for which it was chosen. Similarly, every model will have shortcomings, and personal views concerning the balance between the strengths and weaknesses of a model will vary enormously. Some of the more obvious examples of this are reflected in the long standing, yet unresolved, debates regarding the value of using induced versus spontaneous tumors; the merits of autochthonous tumors versus transplantable tumors (whether of spontaneous origin or induced); and the use of highly antigenic immunogenic tumors versus weakly immunogenic or non-immunogenic (?) tumors in evaluating host immune response to tumors. Similarly, the incidence of spontaneous tumors arising in specific cell types varies significantly in man and animals. Also, the risk of tumor formation occurring in the same target organ after exposure to carcinogens varies markedly between species and is a major limiting factor in using animals to predict the carcinogenic liability of food additive, chemicals or drugs [16,17].

Notwithstanding the complexity of these unresolved issues, several important conceptual and technical advances have been made in the development of metastatic tumor models that must be accommodated in experimental strategies that seek to correlate changes in tumor cell surface properties with the ability to metastasize. First, it is necessary to study metastatic tumors. This may seem to be stating the obvious to the point of absurdity, yet it is still not uncommon, though far less frequent than a few years ago, to read papers in which the stated rationale is to study "malignancy" or "metastatic disease", yet a non-metastasizing tumor is being used. In addition to this fundamental requirement, it is also helpful if the tumor selected for study displays all or most of the following properties: its origin and passage history is known in detail; it can be transplanted in syngeneic hosts or immune-deficient animals to produce a reproducible and consistent pattern of metastatic disease in defined organ(s); the plating efficiency of the cells is sufficient to allow routine recovery of tumor cells from tumors in vivo; the phenotypic properties expressed by defined subpopulations of tumor cells in vitro persists in vivo and viceversa; assay procedures are available to quantify the metastatic burden; and for the development of therapeutic agents it is useful if the tumor exhibits a similar response profile to human tumors of comparable histologic origin.

Second, with the demonstration that the cellular composition of most, if not all, malignant tumors is heterogeneous, and that only certain subpopulations of tumor cells express metastatic properties, perhaps in a transient fashion, it is now almost obligatory to isolate representative examples of the constituent cellular subpopulations for detailed comparison [1,2,8,11,18-20].

TUMOR CELL HETEROGENEITY AND CORRELATION OF CELL SURFACE CHANGES
WITH METASTATIC BEHAVIOR

The knowledge that all human and animal tumor cell populations
studied to date have been found to contain phenotypically different
subpopulations of cells has important implications for experimental
efforts to define the cellular and subcellular properties that
correlate with complex, multifactorial behavioral traits such as
metastasis. As discussed at length elsewhere, the coexistence within
the same tumor (or tumor cell line) of subpopulations of cells with
different metastatic properties dictates that the use of heterogeneous
populations containing both non-metastatic and metastatic cells
cannot be interpreted as yielding information about the metastatic
subpopulations per se [2,6,11,21]. If, for example, the metastatic
cell subpopulations are in the minority, analysis of the entire cell
population may provide little or no insight into the cell surface
properties that are unique to metastatic cells since the probability
is high that these would be obscured or lost in the "background
noise" imposed by the majority fraction of non-metastatic cells [2].
Consequently, characterization of the surface properties of metastatic tumor cell subpopulations requires that either such subpopulations are present as the overwhelming majority in a heterogeneous
cell population or that individual subpopulations be isolated by
cloning to generate homogeneous cell preparations for comparison
with non-metastatic clones isolated from the same parent cell
population. In the former strategy, selection pressures are imposed
to favor the metastatic subpopulations, with resulting enrichment of
their contribution to the total cell population [2]. Conversely,
negative selection pressures can be applied to deplete the fraction
of metastatic cells [2]. It is important to emphasize, however, that
tumor cell preparations subjected to selection pressures to enrich
or deplete specific subpopulations of cells are still heterogeneous
and that non-metastatic subpopulations are still likely to be present.

A more satisfactory approach for addressing the problems posed
by tumor cell heterogeneity involves cloning of tumor cells from a
heterogeneous cell population to create a panel of clonal populations
whose phenotypes can be compared [2]. In this strategy the aim is to
isolate a sufficient number of clones so that clones with the
following properties can be compared in order to identify surface
properties that are invariant features of metastatic clones but
which are absent from non-metastatic clones:

tumorigenic, non-invasive, non-metastatic clones $(T^+1^-M^-)$
versus
tumorigenic, invasive, non-metastatic clones $(T^+1^+M^-)$
versus
tumorigenic, invasive, metastatic $(T^+1^+M^+)$

Although this approach is conceptually straightforward, it is technically demanding and labor-intensive because of the need to isolate and examine a large number of clones from the same parent tumor cell population. Furthermore, as discussed in detail elsewhere [2], reliable correlation of a specific cell surface change assayed in vitro with the metastatic properties of cells in vivo will require additional experiments to confirm that the same surface change is expressed by the metastatic cells in vivo and in cells recovered from metastatic lesions and cultivated in vitro surface properties [22,23].

Since cell cloning offers the most direct and effective method for studying variation in the surface properties of different cells residing in the same population, it is helpful if the tumor selected for study has a reasonably high cloning efficiency. Furthermore, the use of defined media to facilitate the cloning and the recovery of the more fastidious tumor cell clones merits attention. These requirements are of particular relevance to studies with human tumor cells and a substantial effort will be required to solve these problems before detailed analyses of the clonal composition of tumors of the kind already completed for a few animal tumors can become routine practice for human tumors. Identification of clonal variation among cells present in a specific tumor also demands that a panel of stable phenotypic markers, including surface markers that are suitable both for in vitro and in vivo analyses be available to allow reliable identification of individual clones within heterogenous populations and to follow clonal lineage[22,23]. Ideally, the panel should include a series of karyotypic, immunologic and biochemical markers which can be assayed in reproducible fashion in tumor cells both in vitro and in vivo [15]. As emphasized earlier, few panels of this kind are presently available. Also, for experimental purposes it is acceptable to engineer the introduction of specific markers into individual subpopulations using gene transfer methods, mutagenesis or treatments that induce significant phenotypic change such as treatment with drugs that alter DNA methylation patterns [15]. In creating a panel of phenotypic markers it is essential, however, to determine the (in)stability of such markers in clonal tumor cell populations when maintained under different conditions in vitro and in vivo (see below).

In comparing the surface properties of metastatic and non-metastatic clones isolated from the same parent cell population it is necessary to confirm that the metastatic properties of the clones are stable. This is assessed by repeated subcloning at intervals of only a few weeks to examine whether the metastatic behavior of the original clone and progeny subclones are maintained or if variant subclones whose metastatic properties differ significantly from the original clone emerge to create a heterogeneous cell population. Studies on several animal tumors have revealed that the metastatic properties of certain clones may be highly unstable and

that subclones with altered metastatic behavior emerge rapidly,
generating a heterogeneous cell population [2,11,20].

If clones from the tumor being studied undergo rapid phenotypic
"drift", frequent subcloning will be needed to ensure that the clonal
populations being compared are homogeneous and that repeat experiments are done using replicate subclones that display identical
metastatic properties to the original clones. It is also informative to obtain data on the rate of phenotypic "drift". If no
significant changes in metastatic behavior, or other phenotypic
properties, occur, or occur only over a period of several months
rather than weeks, in vitro analyses of specific tumor cell properties can be undertaken in the reasonable belief that the metastatic
properties of the cells being assayed are still comparable to samples
of the same clone assayed a few weeks before or after the experiment.
If, however, the rate of drift in metastatic properties is rapid due
to the emergence of variant subclones with altered metastatic abilit-
ies, the experimental requirements for correlating cell surface
alterations assayed in vitro with metastatic behavior in vivo become
far more stringent. For clones that show rapid drift in metastatic
behavior (or other properties), the correlation of a specific surface
property assayed in vitro with metastatic capacity in vivo will require that the in vitro and in vivo assays be conducted synchronously
using replicate cell preparations assayed at the shortest interval
after cloning consistent with generating sufficient cells for the
assays [24]. Unless such precautions are taken, the in vitro assay of
a cell surface property today cannot be correlated with the metastatic behavior of the same "clone" measured a few weeks earlier or
later, because rapid formation of variant subclones will have rendered the "clone" heterogeneous. Consequently, for tumor cell clones
that exhibit unstable metastatic properties during serial passage
these more demanding synchronous assay protocols must be adopted as
routine practice, unless experimental conditions can be developed
to eliminate the drift in metastatic properties [2].

The liability of metastatic properties and many other phenotypic
traits in tumor cell clones during serial passage in vitro and in
vivo poses formidable logistical problems in designing experiments
to correlate cell surface changes with metastatic competence or,
at the next level of analytical detail, correlation with specific
steps in the metastatic process, such as invasion, tumor cell
survival in the circulation and arrest. Although polyclonal cell
lines are relatively stable under similar conditions the disadvant-
ages of employing these heterogenous populations have been mentioned
already. Consequently, to ensure that in vitro assays of biochemical
or molecular traits can be related to the in vivo behavior of the
cells being studied, correlative experiments are performed synchronously using replicate tumor cell populations on the same day with
an addition flask harvested for storage as a reference stock [24].

By far the most perplexing problem in performing experiments of this type is the time required for expanding any given clone to generate sufficient cells for biochemical and biological analysis. There is little guarantee that during this relatively lengthy procedure (6 to 10 wks) that the "clone" will not be rendered heterogeneous by rapid formation of new variant subclones with altered biological and biochemical phenotypic traits. To minimize complications attributable to phenotypic destabilization, the following precautions can be taken: all tumor cell clones should be used as soon as possible following their initial isolation; biochemical or other in vitro analyses of cell surface properties are performed on a panel of clones so that atypical information derived from a single clone that has undergone rapid phenotypic "destabilization" will not have a disproportionate effect on the interpretation of the data; and at the time of the final in vivo and in vitro analyses, clones should be subcloned to evaluate the homogeneity of the parent clone, with approximately 10 to 15 subclones being tested from each parent clone. Although the last procedure is the most informative in evaluating the phenotypic stability of the original clone, it is too tedious and time consuming to be performed on a routine basis and consequently only a minority of clones in a given screen can be investigated in this way. The dilema of phenotypic drift can never be totally circumvented, but the above precautions are at least sufficient to ensure that the majority of tumor cell clones being studied represent homogeneous cell populations that can be legitimately compared with each other. This set of precautions is also adequate to detect phenotypic 'drift' in clones. For example, the ability of certain B16 melanoma clones to accumulate cyclic AMP when challenged with activators of adenylate cyclase (forskolin and MSH) has been shown to correlate positively with their metastatic performance [25]. This conclusion was based on a study of over 30 B16 melanoma clones. In pursuing this observation, two clones, F1-C14 and F1-C16, were selected for detailed biochemical investigation. During cultivation in vitro, the response of these clones to forskolin and MSH challenge was found to alter radically and was accompanied by a concomitant change in metastatic properties [25]. Phenotypic 'drift' in these clones was also confirmed by subcloning experiments which demonstrated the emergence of phenotypically heterogenous subclones in both "clones" [25]. In contrast it should also be pointed out that unpublished observations of Lester et al have shown that a number of other B16 clones isolated and expanded in culture at the same time as F1-C14 and F1-C16 have displayed remarkable phenotypic stability in that their metastatic capacity and hormonal responsiveness have remained unaltered following continuous propogation in vitro (and in vivo) for several months.

If the cell surface change being studied is relevant to the expression of the metastatic phenotype, it would be reasonable to predict that metastases formed in vivo should express the same alteration. It is therefore informative to examine expression of

the property in question in biopsies both from the primary tumor
implant, and more important, individual lung metastases, either
assayed directly after excision from the animal or following recovery
of cells from the lesion and establishment in culture [22,23]. Demonstration of an association between a specific cell surface alteration
and metastatic performance in other tumor systems of similar or dis-
similar histologic origins would then further strengthen the proposed
association.

OBLIGATORY AND NON-OBLIGATORY PHENOTYPES: DEMONSTRATING CAUSALITY
BETWEEN PHENOTYPIC CHANGE AND METASTATIC BEHAVIOR

Even when the rigorous methods outlined in the previous section
are used, further technical complexity is encountered in attempting
to define which of the many altered properties exhibited by tumor
cells are obligatory for expression of tumorigenicity or metastasis
(obligatory phenotypes) versus traits that may confer valuable
adaptive advantage on tumor cells, and facilitate their survival
in the face of potentially destructive selection pressures imposed
by the host or therapy, but which are not absolutely necessary for
expression of tumorigenic or metastatic behavior (non-obligatory
phenotypes). Stated another way, the task facing investigators
interested in defining the molecular basis of metastatic behavior
is how to distinguish the various obligatory phenotypic traits that
are presumed to be expressed by all metastatic cells (i.e. homogeneous traits) against the "background noise" of extensive phenotypic heterogeneity created by variable expression of non-obligatory
traits that are, in the strictest sense, irrelevant to the expression
of tumorigenic of metastatic behavior. This assumes, however, that
all of the metastatic tumor cell subpopulations present in different
tumors of common histologic origin will exhibit the same panel of
obligatory phenotypic changes. It is perhaps not unreasonable to
assume that at least some obligatory phenotypes will be shared and
should this be detected as an invariant feature of metastatic cells,
and some of these might be shared with metastatic tumor cell subpopulations in neoplasms of different histologic origins arising in
tissues with the same germ cell layer origins. However, as outlined
below, metastatic behavior might also arise via the expression of a
set of phenotypic changes which collectively confer metastatic
competence (i.e. complementing phenotypes) but which cannot induce
this behavioral change in a cell if expressed individually. Such
phenotypes will only be detected by analysis of the expression of
multiple phenotypes in a large number of tumor cell clones. The
technical demands of this exercise do not require emphasis and it
is probably impossible using today's methods. Also, the identification of complementing phenotypes requires that all of the phenotypes
involved are already identified. If complementation involves a
trait that has yet to be detected, the correlation of the other
complementing traits to metastatic behavior will be impossible.
Thus, "known" traits could be coexpressed in clones that are non-

metastatic because they lack the yet "undiscovered" complementing
trait. Coexpression of the "known" traits in metastatic clones
that express the yet "undiscovered" trait would then create the
risk of drawing the erroneous conclusion that the known traits have
no relevance to metastasis. The scale of the analysis needed to
identify complementing phenotypes will increase on a daunting scale
as the number of traits involved increases.

Theoretically, cell surface changes and other phenotypic alterations found in tumor cells can be assigned to a number of categories based on their functional relevance.

Obligatory Phenotypes

As outlined above, these represent changes in cell surface
properties or other cellular traits that are essential for full
expression of a multi-factorial trait such as tumorigenicity or
metastasis. Cells lacking the ability to express an obligatory
metastatic phenotype would not be metastatic. However, expression
of an obligatory phenotype need not be a permanent feature, and
might only be detectable at certain specific stages of the metastatic
process [2]. The concept of temporal variation in the expression of
metastatic competence is consistent with the idea of "transient
metastatic compartments" as proposed by Weiss [26]. According to this
hypothesis, all tumor cells within the primary neoplasm have the
potential to metastasize but the expression of this behavior in an
individual tumor cell is only temporary. Tumor cells are envisaged
as moving asynchronously through a series of functional states,
referred to as compartments, by Weiss [26], and thus constantly acquir-
ing, expressing and losing the capacity to metastasize under these
circumstances. The expression of obligatory molecular traits associated with the metastatic phenotype would also be transient thus
enormously complicating their identification and characterization.

Obligatory properties can be subdivided into two groups:
"threshold" phenotypes and "incremental" phenotypes. A "threshold"
property is one that is expressed by all metastatic cells (for a
given tumor type), but which displays little or no quantitative
change with variation in metastatic proficiency. In other words
the successfully metastatic cell need not excel at completing all
of the steps in the metastatic process. It need only display
"minimal competence". For example, the most invasive tumor cells
(as judged, for example, by secretion of degradative enzymes) need
not necessarily be the most metastatic. A metastatic cell needs
only to produce sufficient proteolytic enzymes to achieve breakdown
of the extracellular matrix to provide access to the vasculature.
The concept of "minimal competence" cautions against expecting a
linear or simplistic relationship between a given biochemical trait
and metastatic performance. Rather, metastatic tumor cells may
require only a critical "threshold" of functional competence in the

properties required to complete any particular step in the metastatic process. In contrast to threshold phenotypes, "incremental" pheno-types are also expressed by all metastatic cells but show significant quantitative variation (positive or negative) in association with gradations in metastatic proficiency. For example, the ability of circulating metastatic tumor cells to respond to local growth factors may be related directly to the number of cell surface receptors for the mitogen and the rate of proliferation of individual metastases will be a direct reflection of receptor density on the cell surface.

Non-Obligatory Phenotypes

Secondary (Advantageous) Phenotypes

Some of the changes detected in tumor cells may not be oblig-atory for metastasis and can thus be classified as secondary or advantageous phenotypes. Such properties are not required for metastasis but their presence confers significant adaptive advantage upon the metastatic tumor cell in the face of potentially destructive selection pressures mounted by host defences and competition by other clones within the tumor. Examples of this kind might include a greater tolerance of the tumor cell to stressful conditions (e.g. hypoxia, pH fluctuations, nutrient deprivation) and reduced sensit-ivity to recognition and destruction by host defence mechanisms and therapeutic agents.

Predisposing Phenotypes

Certain phenotypic changes detected in tumor cells may not be essential for metastasis but their presence may enhance the risk and/or frequency with which obligatory metastatic properties are generated. Error-prone DNA polymerases and gene transpositions might represent examples of "predisposing" phenotypes.

Complementing Phenotypes

The final category of phenotypic changes that may occur in tumor cells represent traits that if expressed singly do not confer metastatic competence but when expressed concomitantly generate metastatic cells. With our current limited level of analytical sophistication, the identification of complementing phenotypes will be extremely difficult to achieve. Neverthless, their existence, although theoretical, would not be unexpected. For example, the ability of a tumor cell to synthesize and secrete its own growth factors (phenotype 1) confers no advantage benefit unless the cell is also capable of synthesizing and expressing the specific cell surface receptor (phenotype 2), with aberrant autocrine regulations occurring as a consequence of these two "complementing" phenotypes. Complementing interactions may also occur between cells. Non-invasive tumor cells may gain access to the circulation, not through their own efforts, but by exploiting the invasive capacity of neighboring tumor cells. However, even though they may lack the properties to invade the circulation in their own right, once intravasation has occurred they may possess all of the traits needed to complete the remaining steps in the metastatic process and thus be able to form metastases. Similarly, a tumor cell displaying surface determinants recognized by host defence mechanisms could be passively carried through the vasculature in a protected manner by passive entrapment within a clump of tumor cells that are not recognised by circulating NK cells. The existence of complementing phenotypes also implies that the isolation of tumor cell populations from metastases which are devoid of a biochemical trait previously considered to be obligatory for metastatic competence does not necessarily negate the importance of the trait in the expression of the malignant phenotype since cellular complementation may have enabled the cells recovered from the lesion to give rise to a metastatis. However, in such cases, the isolated tumor cell population would not be expected to metastasize when reinjected into animals since, by itself, it lacks one or more phenotypic traits essential for successful completion of the metastatic process.

It could be argued that since tumorigenicity and metastasis are complex composite traits that require alterations in a wide variety of cell properties, each of these alterations, including obligatory phenotypes, represent complementing phenotypes. This is true only if metastasis is viewed as a single event. It clearly is not and comprises a series of sequential steps. The ability of a tumor cell to complete any particular step could presumably be shown to result from specific tumor cell properties if suitable assays were available to allow experimental dissection of metastasis into discrete stages. In short, the functional importance of changes in cell surface properties, or other cellular functions, will eventually need to be defined in relation to specific steps in metastasis, rather than the entire metastatic cascade. This requires, of course, that assays are available to assay the competence of tumor cells to complete each step in the metastatic process. Experimental efforts to develop such assays are only just beginning.

When defined in relation to a specific stage in the metastatic process rather than the entire process, the distinction between obligatory and complementing phenotypes is clearer. Expression of an obligatory phenotype by a tumor cell will automatically allow it to complete the step in question. In contrast, the ability of complementing phenotypes to substitute for an obligatory phenotype in allowing a tumor to complete the same step will require that all of the interacting phenotypes be expressed. None of these traits are able to render the cell competent to complete the same steps if expressed in anything other than the full complement and thus cannot be viewed as obligatory phenotypes in their own right.

Finally, irrespective of the type of biochemical trait being sought, a long recognized concern, still unresolved, is the issue of how tumor cell properties examined in vitro faithfully reflect those exhibited by the same cells in vivo. Past failures to reveal biochemical correlates of the metastatic phenotype may be explained in part, by the artificial environment of in vitro culture conditions. Under the relatively quiescent conditions of in vitro culture, and in the absence of host selection pressures, such phenotypes may not be activated and thus go undetected. Biochemically, tumor cell populations with widely differing metastatic capacities may be indistinguishable in vitro. However, following inoculation into animals and exposure to a multiplicity of host stimuli they may display a spectrum of biochemical responses that account for their distinct metastatic capacities. Investigations of metastatic clones derived from the B16 melanoma have confirmed this suspicion and revealed that some biochemical differences between clones with different metastatic capacities are more apparent in cultures that have been exposed to metabolic challenges such as exogenous hormones, growth factors, and stressing agents than unchallenged cultures [25].

EXPERIMENTAL MANIPULATION OF THE METASTATIC PROPERTIES OF TUMOR CELLS: FULFILLING KOCH'S POSTULATES AT THE MOLECULAR LEVEL

To establish a causal relationship between infection by specific micro-organisms and development of disease, the nineteenth century microbiologist Robert Koch formulated a set of experimental requirements that are equally applicable to contemporary efforts to demonstrate causal correlations between specific changes in tumor cells and their behavior in vivo. Koch's postulates required that the micro-organism suspected of causing disease should be found consistently in all cases of the disease, should be recovered consistently from disease lesions and produce the disease when introduced into susceptible hosts. In analyzing causal events in tumor cell behavior, the equivalent of the first two of Koch's postulates would be satisfied by the experimental approaches described in previous sections in which a specific cell surface change is shown to be a consistent feature of: 1) metastatic cells in vitro; 2) cells in metastatic lesions in vivo; and 3) cells recovered from metastases and recultured in vitro. The equivalent of the third of Koch's postulates of causality would be to demonstrate that experimental manipulations that induce loss or acquisition of the surface property of interest are accompanied by loss or gain of metastatic capacity.

The study of microbial disease has the substantial advantage that disease results from the action of a single agent (pathogenic micro-organism) and the demonstration of causality is relatively straightforward (at least to the 20th century pathologist). Disease can be induced by introduction of the organism into a susceptable host and progression of the disease can be arrested by agents that impair microbial function (host immunity, vaccination, antibiotics). In contrast, the suspected multiplicity of phenotypic changes needed to confer metastatic behavior on a cell dictate that it is far more complicated to show whether loss or gain of any specific cellular characteristic affects metastatic behavior. It may be easier for example to induce loss of metastatic ability that acquisition of this behavior. Experimentally-induced loss of a single obligatory phenotypic trait may be sufficient to abrogate the entire metastatic process by rendering a tumor cell incapable of completing just one step in the metastatic process. It is important to emphasize, however, that depending on the mechanism of action of the manipulation responsible for loss of metastatic competence, other phenotype traits that are essential for metastasis may continue to be expressed and it would be erroneous to conclude that their expression in cells that have been rendered non-metastatic invalidates any previously established causal correlations for these traits.

To confer metastatic ability on a tumorigenic but non-metastatic cell by experimental alteration of a single phenotypic trait is likely to be far more difficult. Success would probably require that the change be imposed on a cell that possesses all of the other phenotypic changes needed to metastasize except the property being investigated. A possible experimental strategy in this regard would be to show that experimentally-induced loss of a single (obligatory) phenotype from a metastatic cell eliminates its metastatic capacity and that induction of the same trait in the modified cell restores metastatic behavior. In contrast, conversion of non-neoplastic cells and many tumorigenic, non-metastatic cells into metastatic cells may require simultaneous alterations in a variety of cell properties that are not technically feasible using current techniques.

Our present ignorance of the extent and nature of the phenotypic changes needed to confer metastatic properties on a malignant tumor cell suggests that a more productive approach is to focus experimental questions on how changes in specific surface properties affect the ability of tumor cells to complete specific steps in the metastatic process. By defining the functional importance of a specific property at different stages in the metastatic cascade a catalog of phenotypic changes needed to complete the entire process might eventually be assembled. For example, Poste and Nicolson [27] showed that when plasma membrane vesicles isolated from B16 melanoma cells that metastasized preferentially to the lung were fused with B16 cells that were less efficient in metastasizing to this organ, the "membrane-modified" cells displayed a significantly higher arrest in the lung microcirculation. This experiment suggests that plasma membrane components introduced from the vesicles could alter the organ distribution and arrest patterns of circulating B16 cells.

The powerful new methods for altering cell function that are now emerging as a result of rapid advances in molecular genetics give

significant opportunities for exploring the functional importance
of specific surface determinants at different steps in the metastatic process. For example, although the evidence remains circumstantial, there is a strong belief that the secretion of proteases
by malignant cells is essential for successful metastasis. First,
to abrogate protease activity by using selective enzyme inhibitors
or second, by endowing a tumorigenic non-metastatic cell with protease activity. Theoretically the latter can now be achieved using
molecular genetic techniques. Once cloned, the gene for a particular tissue degradative enzyme (e.g. collagenase, plasminogen
activator) can be introduced into the target tumor cell and its
effect on metastatic properties assayed. In future, similar experiments should be possible for a whole range of tumor cell gene
products implicated in the metastatic process; for example laminin
receptors, adenylate cyclase, enzymes regulating arachadonic acid
metabolism and the various surface determinants whose expression has
been shown to vary with changes in metastatic proficiency [11]. These
techniques offer powerful new approaches for the identification of
specific biochemical properties that may contribute to the expression
of metastatic behavior. Advances in DNA technology also raise the
distant, but nevertheless, intriguing prospect of targeting drugs
not only to the protein products of genes but also to the genes
themselves. This represents an exciting new area in fundamental
research with direct implications for the treatment of metastasis.

Finally, gene transfer techniques are beginning to be used to
study whether transfection of non-metastatic cells with genomic
material from metastatic cells might render them metastatic in a
fashion analogous to the tumorigenic transformation of cells by
oncogenes. Once again, however, the multifactorial nature of the
metastatic phenotype may be an obstacle to detection of "metastatic
genes". Even if such genes exist, expression of metastatic properties may occur only if these elements are introduced into tumor
cells that already possess many of the phenotypic alterations needed
to metastasize and which have accumulated as a result of multiple
changes in genomic expression during serial cultivation of these
cells in vitro or progressive growth in vivo. This may well be the
case since metastatic cells do not appear to be present from the
outset in host tumors and emerge at a later stage in tumor progression; the exact stage differing between neoplasms arising in
different cell types [28].

Even if transfection of metastatic properties does not prove
to be feasible, it is clear that the new tools of molecular genetics,
coupled with equally powerful improvements in techniques and analytical methods in cell biology and immunology, offer substantial
promise for altering the expression and regulation of specific gene
products in normal and neoplastic cells and herald exciting opportunities for manipulating a wide range of cell properties that can
be confidently expected to be of great value in defining the

functional importance of various alterations in cell surface
properties in determining the aberrant behavior of neoplastic cells.

REFERENCES

1. G. Poste and I.J. Fidler, The pathogenesis of cancer metastasis,
   Nature 283: 139-146, (1980).
2. G. Poste, Experimental systems for analysis of the malignant
   phenotype, Cancer Met. Rev. 1: 121-199, (1982).
3. G.L. Nicolson and L. Milas (eds), Cancer Invasion and Metastasis:
   Biologic and Therapeutic Aspects, Raven Press, New York.
4. N. Bruchovsky and J.H. Goldie (eds) Drug and Hormone Resistance
   in Neoplasia, Volume I and II, CRC Press, Boca Raton.
5. A.J.S. Davies and M.J. Crumpton (eds), Experimental Approaches
   to Drug Targeting, Cancer Surveys, Volume I, pp 349-559,
   Oxford University Press, Oxford.
6. I.J. Fidler and G. Poste The cellular heterogeneity of malignant
   neoplasms: implications for adjuvant chemotherapy, Semin.
   Oncol. in press, (1985).
7. S. Hawkes and J.L. Wang (eds) Extracellular Matrix, Academic
   Press, New York, (1982).
8. A.H. Owens Jr., D.S. Coffey and S.B. Baylin (eds), Tumor Cell
   Heterogeneity: Origins and Implications, Academic Press, New
   York, (1983).
9. G. Poste and G.L. Nicolson, Experimental systems for analysis
   of the surface properties of metastatic tumor cells, in:
   Biomembranes, Volume 11, (A. Nowotny, ed), Plenum Press,
   New York, pp 341-364, (1983).
10. F. Bresciani, R.J.B. King, M.E. Lippman, M. Namer and J.P.
    Raynaud (eds) Hormones and Cancer 2, Volume 31, Progress in
    Cancer Research and Therapy, Raven Press, New York, (1984).
11. G.L. Nicolson and G. Poste, Tumor implantation and invasion of
    metastatic sites. Int.Rev.Exp. Pathol., 25: 77-181, (1984).
12. H.B. Hewitt, Counterpoint: Animal tumor models and their
    relevance to human tumor immunology, J. Biol. Resp.Mod.,
    1: 107-119, (1982).
13. A Goldin, Animal models for cancer chemotherapy, in: Cancer
    Chemotherapy, (F.M. Muggia ed.) Martinus Nijhoff, The Hague,
    pp. 65-102, (1983).
14. M.G. Donelli, M. D'Incalci and S. Grattini, Pharmacokinetic
    studies of anticancer drugs in tumor-bearing animals. Cancer
    Treat.Rep. 63: 381-400, (1984).
15. G. Poste and R. Greig, Experimental models for studying the
    pathogenesis and therapy of metastatic disease, in: Mechanisms
    of Metastasis: Potential Therapeutic Implications (K.V. Honn,
    J.D. Crissman, W.E.Powers and B.F. Sloane, eds), Martinus
    Nijhoff, The Hague, (1985).
16. E. Efron, The Apocalytics, Simon and Schuster, New York,
    (1984).

17. National Toxicology Program, Report of the NTP Ad Hoc Panel on Chemical Carcinogenesis Testing and Evaluation, U.S. Department of Health and Human Services, (1984).

18. I.J. Fidler and G. Poste, The heterogeneity of metastatic properties in malignant tumor cells and regulations of the metastatic phenotype, in: Tumor Cell Heterogeneity (A. Owens, D.S. Coffey and S.B. Baylin, eds), Academic Press, New York, pp. 127–145, (1982).

19. G. Poste and R. Greig, On the genesis and regulation of cellular heterogeneity in malignant tumors, Invasion and Metastasis 2: 137–176, (1982).

20. G.H. Heppner, Tumor heterogeneity, Cancer Res. 44: 2259–2265, (1984).

21. G. Poste and R. Greig, The experimental and clinical implications of cellular heterogeneity in malignant tumors. J. Cancer Res. Clin. Oncol., 106: 159–170, (1983).

22. G. Poste, J. Brown, A.E. Tzeng and I. Ziedman, Comparison of the metastatic properties of B16 melanoma clones isolated from cultured cells lines, subcutaneous tumors and individual lung metastases, Cancer Res. 42: 2770–2778.

23. G. Poste, J. Tzeng, J. Doll, R. Greig, D. Rieman and I. Ziedman, Evolution of tumor cell heterogeneity during progressive growth of individual lung metastasis, Proc. Nat. Acad. Sci., U.S.A. 79: 6574–6578, (1982).

24. R.G. Greig, L. Caltabiano, R. Reid Jr., J. Feild and G. Poste, Heterogeneity of protein phosphorylation in metastatic variants of B16 melanoma, Cancer Res. 42: 6057–6065, (1983).

25. J.R. Sheppard, T.P. Koestler, S.P. Corwin, C. Buscarino, J. Doll, B. Lester, R.G. Greig and G. Poste, Experimental metastasis correlates with cyclic AMP accumulation in B16 melanoma clones, Nature 308: 544–547, (1984).

26. L. Weiss, Metastases: differences between cancer cells in primary and secondary tumors, in: Pathobiology Annual, Volume 10, (H. Loachin, ed.), Raven Press, New York, pp. 51–81, (1980).

27. G. Poste and G.L. Nicolson, Arrest and metastasis of blood-borne tumor cells are modified by fusion of plasma membrane vesicles from highly metastatic cells, Proc. Nat. Acad. Sci., U.S.A. 77: 399–403, (1980).

28. E.V. Sugarbaker, Cancer metastasis: a product of tumor-host interactions, Curr. Problem Cancer, III: 3–59, (1979).

INTERACTION OF PLATELETS, TUMOR CELLS AND

SUBENDOTHELIAL SURFACES

G. A. Jamieson, Eva Bastida and Antonio Ordinas

American Red Cross Laboratories, Bethesda
Md., USA 20814;*Hospital Clinico
Barcelona, Spain+

INTRODUCTION

The metastatic dissemination of tumour cells in the circulation appears to involve at least four distinct stages [1]: (i) the release of cells from the primary tumour; (ii) transportation within the circulation; (iii) attachment to a vessel wall at a distant site; and (iv) extravasation and subsequent proliferation of the secondary tumour. There is considerable evidence that platelets may have a major role in the last three of these stages; namely, the transportation of tumor cells within the circulation, their subsequent attachment to the vessel wall, and the growth of the secondary tumour [2]. In fact, the possible role of platelets in the metastatic process was apparently first recognised from the histological observations of Billroth in 1878 and Schmidt in 1903 on the association of tumor cells with platelet aggregates and thrombi during postmortem examinations. The preferential formation of metastases has also been noted at sites of surgical intervention after removal of the primary tumor [1,3], and at sites of vessel damage.

The mechanism by which platelets enhance metastasis is not known but may involve (i) physical shielding of tumor cells within the platelet embolus from destruction by cytotoxic macrophages; (ii) greater ease of arrest of the platelet-tumor cell emboli in the microvasculature possibly resulting in occlusion, distal hyposia and endothelial damage, and (iii) enhanced extravasation and tumor cell growth as a result of the secretion of glycosidases (heparinase), platelet-derived growth factors and chemotactic factors: in this connection the structural homologies between platelet-derived growth factor and the transforming protein of simian sarcoma virus [5,6] are particularly exciting.

175

In 1962, the Gasics observed that injection of neuraminidase reduced the formation of tumor cell colonies in the lungs of mice injected with mammary adenocarcinoma TA3 cells [7] and in 1968 they made the important quantitative observation that the number of lung tumor cell colonies was reduced in proportion to the degree of thrombocytopenia [8].

TUMOR CELL-INDUCED PLATELET AGGREGATION

Much of the information so far obtained about cancer-platelet relastions in vivo has used laboratory models of tumor growth such as hematogenous dissemination after intravenous injection in mice of large amounts of tumor cells. These models are inherently difficult to work with and, under these experimental conditions, it is not possible to study the early steps of tumor metastasis, for example, the phenomena of cell-cell interactions, tumor cell micro-vesiculation or adherence of tumor cells on vascular structures. For these reasons, many laboratories have examined the ability of tumor cells to induce platelet aggregation. For example, Gasic and co-workers showed that a wide variety of tumor tissues or cultured tumor cells could induce varying degrees of platelet aggregation in heparinized (but not citrated) plasma [9] and similar observations have subsequently been made in numerous other laboratories.

However, the mechanism by which tumor cells activate platelets is not clearly understood. One proposal is that tumor cells induce the release of platelet ADP in an undefined manner [9-11]. A second mechanism suggests the involvement of thrombin since tumor cell-induced platelet aggregation can be blocked by hirudin [12,13] or by dansyl arginine N-(3-ethyl-1,5-pentanediyl)amide (DAPA) [14], and involves the deposition of fibrin [15]. A third possible mechanism was indicated by the fact that aggregating material present in urea extracts of SV40-3T3 fibroblasts appears to require the presence of both complement, and a heat-stable plasma co-factor [16].

Whatever the mechanism, the tumor cell surface appears to play an important role. Platelet activation has been correlated with the surface sialic acid content and metastatic potential of rat renal sarcoma sublines [17] and membrane microvesicles shed from the tumor cell surface can themselves induce platelet aggregation [18]. The latter observation parallels the recent findings that such micro-vesicles have procoagulant activity [19,20] and our own recent results showing that these two activities are causally related in certain cases [21].

A major aspect of our own approach has been to utilize homo-logous systems of human platelets and tumour cell lines of human origin as a way of possibly resolving conflicting reports from different laboratories on tumor cell-induced platelet aggregation carried out in heterologous systems (for example [22,23]). These discrepancies may arise because of species variation in procoagulant effects in heterologous systems, a fact long known [24] but recently rediscovered [20,25].

We have previously shown that normal individuals could be differentiated into responders and non-responders based on the aggregation response of their platelets to individual tumor cell lines at specific cell concentrations [12]. Using responders, we have studied the mechanism of tumor cell-induced platelet aggregation in the human cell lines identified in Table 1. It has long been known that normal cells do not induce platelet aggregation and that cell lines derived from tumors, or from virally-transformed lines, are required [9,18].

The mechanisms of aggregation we have observed appear to fall into two major groups. Aggregation in the larger group, comprising four of the nine cell lines examined, was inhibited by apyrase, by phosphoenol pyruvate/pyruvate kinase and by phospho-mechanism. ADP-dependent aggregation also occurred with the Hut-20 line from an anaplastic murine tumour [12]. Aggregation in the second group, comprising three of the cell lines, was inhibited by hirudin, by heparin at high concentrations and by the specific thrombin inhibitor dansylarginine-N-(3-ethyl-1,5-pentanediyl)amide (DAPA), but not by the ADP-degrading enzymes, defining this as a thrombin-dependent mechanism. Within this thrombin-dependent group, a sub-classification was possible since only Hut-28 was inhibited by phospholipase D. Phospholipase C had no effect with any of the cell lines examined while both phospholipase A2 and lysolecithin inhibited platelet aggregation in every case. Since platelet aggregating factor is destroyed by phospholipases A2, C and D [26] these results show that it is not involved in platelet aggregation induced by these human tumor cell lines. Two of the tumor cell lines, Hut 23 and A549, did not induce platelet aggregation.

Table 1.   Mechanism of Aggregation by Human Tumor Cell Lines

Line	Source	Mechanism
Hut-23	Poorly differentiated adenocarcinoma	No aggregation
Hut-28	Mesothelioma	Thrombin-dependent
A-549	Epithelial lung carcinoma	No aggregation
U87MG	Glioblastoma	Thrombin-dependent
SKBR3	Adenocarcinoma of the breast	ADP-dependent
SKNMC	Neuroblastoma	ADP-dependent
HT-144	Melanoma	ADP-dependent
HT-29	Adenocarcinoma of the colon	ADP-dependent
HL-60	Promyelocytic leukemia	Thrombin-dependent

PERFUSION STUDIES

Studies utilizing tumor-cell induced platelet aggregation do
not take into account the rheological phenomena that occur in the
in vivo situations and which may regulate the interaction of tumor
cells with blood and with vascular structures.

We have recently described the adaptation of the Baumgartner
perfusion system to the study of the interaction of human tumor
cells with blood elements and vascular subendothelium [27]. This
annular perfusion system (Figure 1) was first described by Baumgartner [28] and it has been extensively used to study hemostatic
disorders in platelet defects such as Bernard Soulier syndrome or
Glanzmann's thrombasthenia [29] as well as plasma deficiencies like
the afibrinogenemias or von Willebrand disease [30]. The use of this
perfusion system allows the utilization of vascular material from
different sources. The everted vessel segment that is placed on the
plastic rod can be human renal artery [31], human umbilican cord [32] or
rabbit abdominal aorta [33]. Any of these vascular segments has the
required diameter to provide the correct rheological conditions so
that the perfusion system gives rise to shear rates corresponding
to those in the human vasculature.

Using the perfusion apparatus, we have carried out extensive
studies with the Hut-20 (ADP-dependent) cell line derived from an
anaplastic murine tumor [27]. These cells were included in platelet
thrombi deposited on vascular subendothelium in perfusion experiments with heparinized human blood. Extremely large platelet-tumor
cell thrombi were found at the vascular surface in Hut 20 perfusions
using vessel segments which had been treated with chymotrypsin.
These large heterogeneous thrombi perturbed blood flow through the
system and entrapped both erythrocytes and white cells. When Hut 20
cells were labelled with ^{125}I-deoxyuridine and perfused in whole
blood at a concentration of 3.7×10^5/ml, tumor cell incorporation
into platelet-tumor cell thrombi on chymotrypsinized segments yielded
about 330,000 cpm/mg of vascular tissue. However, this value was

Figure 1.   The Baumgartner perfusion chamber. De-endothelialized
            blood vessel segments are everted and mounted on
            the central rod. Perfusion with heparinized whole
            blood is carried out for 10 mins. in the presence of
            tumor cells or microvesicles. Following this, the
            segments are removed, fixed and stained for morpho-
            metric evaluation.

reduced some two orders of magnitude by the inclusion of PGE_1
(1ng/ml of perfusing blood; 2.8 uM) in parallel samples. Aspirin
at 100 uM reduced tumor cell-dependent platelet aggregation but
did not decrease the platelet-dependent deposition of radiolabelled
Hut 20 cells on vascular subendothelium, suggesting that the release
reaction may not be of major significance in this interaction.
Tumor cell-induced platelet aggregation was not observed in a perfusion experiment using blood from a patient with severe von
Willebrand disease. However, addition of 0.1 vol. of ABO-compatible,
heterologous plasma as a source of factor VIII to the von Willebrand
blood sample restored the platelet-dependent deposition of radio-
labelled tumor cells to control values.

These perfusion studies have been extended and confirmed using
the tumor cell lines SKNMC (ADP-dependent) and U87MG (thrombin
dependent). We have found that the thrombin dependent line was
more thrombogenic than the ADP-dependent line and that the thrombogenicity of both lines was increased when perfusion was carried out
on aortic segments which had been digested with chymotrypsin.

ROLE OF DIVALENT CATIONS

We have addressed in detail the question of the role of divalent
cations in the interaction of platelets with tumor cells (Bastida
et al, submitted for publication). Previously it had been generally
accepted that tumor cell-induced platelet aggregation could be observed only in heparinized plasma but a recent report suggested that
the inhibitory effect of citrate was due to chelation of Mg^{++} rather
than Ca^{++} in the aggregation of mouse and rabbit platelets by two
mouse tumor cell lines [11] and in other systems [23,34]. In our homologous human system we have found that Ca^{++} is obligatory for platelet aggregation induced by either a thrombin-dependent cell line
(U87MG) or by an ADP-dependent cell line (SKNMC). With the U87MG
(thrombin-dependent) cell line, the use of a higher molecular weight
heparin fraction with strong antithrombin III activity caused a
pronounced prolongation of the lag time to the onset of aggregation
in comparison with a lower molecular weight heparin fraction with
higher activity against Factor Xa: maximal aggregation, once initiated, was at the same rate and extent in each case. Using the Baumgartner perfusion apparatus, we found that thrombogenesis was much
greater with U87MG (thrombin-dependent) cells than with SKNMC (ADP-
dependent) cells but interaction with subendothelium was reduced
to control levels in the presence of citrate.

These results show that chelation of Ca^{++} ions by citrate
inhibits both platelet aggregation and interaction with subendothelium with both ADP- and thrombin-dependent mechanisms. Thus,
Ca^{++} has an essential role in all forms of platelet-tumor cell
interaction in these homologous human systems.

ROLE OF MICROVESICLES

We have recently examined the formation of microvesicles by
the U87MG and HL-60 (thrombin-dependent) lines and by the SKNMC
(ADP-dependent) line. Each of these lines appears able to elaborate
microvesicles which can induce platelet aggregation in a manner
identical to that of the intact cells. The activity of the microvesicles from the thrombin-dependent lines were blocked by DAPA and
by antitissue factor antibody [21].

The results suggest that both procoagulant and proaggregating
activities are causally related through the presence of tissue
factor in the microvesicles: The production of minute quantities
of thrombin by the extrinsic pathway causes platelet activation
and the further rapid generation of thrombin by both the intrinsic
and extrinsic pathways leading to platelet aggregation and coagulation. Previous reports of tumor cell microvesicles showing
platelet aggregatory activity [18] and procoagulant activity [20] may,
in some cases, be individual aspects of the same phenomenon.

The Baumgartner perfusion apparatus has been used to compare
the interaction of platelets, tumor cells and microvesicles from
the SKNMC and U87MG lines with subendothelium using rabbit aortic
segments before and after digestion with chymotrypsin (Bastida et
al, submitted for publication). In no case was there any difference
between intact cells and microvesicles in their ability to induce
platelet adhesion to subendothelium. However, microvesicles from
U87MG cells actually formed larger thrombi on undigested segments
than did the intact cells ($p<0.05$). On chymotrypsin-digested
vessels the thrombus size was about 7-fold larger for microvesicles
than on undigested segments and about 15-fold larger for U87MG cells,
and the microvesicles continued to form larger thrombi than did the
intact cells ($p<0.01$). Intact SKNMC cells and microvesicles gave
the same size of thrombus on undigested segments. On digested
segments the thrombus size was about 5-fold larger and was significantly larger for microvesicles than for SKNMC cells ($p<0.01$).

The results show that microvesicles stimulate simiarl, or
even greater, interactions between platelets and microvesicles
than do intact cells with thrombin-dependent or ADP-dependent
activities and further support the concept that intact tumor cells
may not be necessary for the thromboembolic complications of
malignancy.

CONCLUSIONS

The possible roles of platelets in metastases involves a
variety of surface interactions including platelet-to-subendo-
thelium in the adhesion phase, platelet-to-platelet during thrombogenesis and platelet-to-tumor cell or microvesicle surface in

direct interaction. Humoral components of the plasma are also
involved since these interactions do not occur using washed or
gel-filtered platelets. Further complexity arises from the fact
that different cell types utilize either ADP-dependent or thrombin-
dependent mechanisms. An understanding of the complexity of these
mechanisms is necessary in consideration of drug design for possibly
interrupting specific steps in the metastatic sequence.

ACKNOWLEDGEMENTS

     The authors' work presented in this review has been supported
by USPHS grants CA30538 and RR 05737, and grant 962 from the
Spanish Comision Asessora de Investigacion Scientifica y Tecnica.

REFERENCES

1.   E.V. Sugarbaker and A.S. Ketcham, Mechanism and prevention
     of cancer dissemination: An overview. Sem. Oncol. 4:19 (1977)
2.   E. Roos and K.P. Dingemans, Mechanism of metastasis. Biochim.
     Biophys. Acta. 560:315 (1979)
3.   I.J. Fidler and G.L. Nicolson, Immunobiology of the experiment-
     al Metastatic melanoma. Cancer Biology Review 2,171-178.(1981)
4.   B.A. Warren, Platelet-tumor cell interaction. Morphological
     studies, in de Gaetano, G and Garatani S., eds. Platelets:
     A multidisciplinary approach, New Ork, Raven Press pp.
     427-440 ((1978)
5.   T.F. Deuel, J.S. Huang, S.S. Huang, P. Stroobant and M.D.
     Waterfield, Expression of platelet-derived growth factor-
     like protein in simian virus transformed cells. Science
     221:1348 (1983).
6.   M.D. Waterfield, G.T. Scrace, N. Whittle, P. Stroobant,
     A. Johnsson, A. Wasteson, B. Westermark, C-H Heldin,
     J.S. Huang and T.F. Deuel, Platelet-derived growth factor
     is structurally related to the putative transforming
     protein p28^{sis} of simian sarcoma virus. Nature 304:35
     (1983).
7.   G.J. Gasic and T.B. Gasic, Removal of sialic acid from the
     cell coat in tumor cells and vascular endothelium and its
     effects on metastasis. Proc. Natl. Acad. Sci. (USA) 48:1172
     (1962).
8.   G.J. Gasic, T.B. Gasic and C.C. Stewart, Antimetastatic effects
     associated with platelet reduction. Proc. Natl. Acad. Sci.
     (USA), 61:46, (1968).
9.   G.J. Gasic, P.A.G. Kock, B. Hsu, T.B. Gasic and S. Niewiarowski,
     Thromogenic activity of mouse and human tumor: Effects on
     platelets, coagulation and fibrinolysis, and possible sig-
     nificance for metastases. S. Krebsforsch. 86:263, (1976)
10. R. Holme, R. Oftebro and T. Hovig, In vitro interaction between
     cultured cells and blood platelets. Thromb. Haemostas.
     40:89, (1978).

11. Y. Hara, S. Steiner and M.G. Baldini. Characterization of the
     platelet aggregating activity of tumor cells. Cancer Res.
     40:1217, (1980).
12. E. Bastida, A. Ordinas and G.A. Jamieson. Idiosyncratic
     platelet responses to human tumor cells. Nature 291:661,
     (1981).
13. E. Bastida, A. Ordinas, S.L. Giardina and G.A. Jamieson.
     Differentation of platelet aggregating effects of human
     tumor cell lines based on inhibition studies with apyrase,
     hirudin and phospholipase. Cancer Res. 42:4348. (1982).
14. E.P. Pearlstein, C. Ambrogio, G.J. Gasic and S. Karpatkin,
     Inhibition of platelet-aggregating activity of two human
     adenocarcinomas of the colon and an anaplastic murine
     tumor with a specific thrombin inhibitor, dansylarginine
     N-(3-ethyl-1,5-pentanediyl)amide. Cancer Res. 41:4533.
     (1981).
15. H. Al-Mondhiry, V. McGarvey and K. Leitzel. Interaction of
     human tumor cells with human platelets and the coagulation
     system. Thromb. Haemostas. 50:726. (1983).
16. S. Karpatkin, A. Smerling and E.P. Pearlstein. Plasma
     requirement for the aggregation of rabbit platelets by an
     aggregating material derived from SV40-transformed 3T3
     fibroblasts. J. Lab. Clin. Med. 96:994.
17. E.P. Pearlstein, P.L. Salk, G. Yokeeswaran and S. Karpatkin,
     Correlation between spontaneous metastatic potential,
     platelet-aggregating activity of cell extracts, and cell
     surface sialylation in 10 metastatic-variant derivatives
     of a rat renal sarcoma line. Proc. Natl. Acad. Sci. (USA)
     77:4336. (1980).
18. G.J. Gasic, D. Boettiger, J.L. Catalfamo, T.B. Gasic and
     G.J. Stewart, Aggregation of platelets and cell membrane
     vesiculation by rat cells transformed in vitro by Rous
     Sarcoma virus, Cancer Res. 38:2050, (1978).
19. H.F. Dvorak, S.C. Quay, N.S. Orenstein, A.M. Dvorak, P. Hahn,
     A.N. Bitzer and A.C. Carvalho, Tumor shedding and coag-
     ulation. Science. 212:923, (1981).
20. H.F. Dvorak, L. von de Water, A.N. Bitzer, A.M. Dvorak, D.
     Andersson, S. Harvey, R. Bach, G.L. Davis, V. DeWolf and
     A. Carvalho. Procoagulant activity associated with plasma
     membrane vesicles shed by cultured tumor cells. Cancer
     Res. 43:4345. (1983).
21. E. Bastida, A. Ordinas, G. Escolar and G.A. Jamieson, Tissue
     factor in microvesicles shed from U87MG human glioblastoma
     cells induces coagulation, platelet aggregation and thrombo-
     genesis, Blood 64:177. (1984).
22. N. Tanaka, S. Ashida, A. Tohga and H. Ogawa, Platelet aggreg-
     ating activities of metastasizing tumor cells. Invasion
     Metastasis. 2:289.
23. P.M. Evans and F.P. Cowie, A species difference in platelet
     aggregation induced by tumor cells. Cell Biol. Int. Rep.
     7:771. (1983)

24. A.J. Quick, On various properties of thromboplastin (aqueous tissue extracts). Am. J. Physiol. 114:282. (1935).
25. F. Rickles and R.L. Edwards, Activation of blood coagulation in cancer: Trousseau's syndrome revisited. Blood 62:14. (1983).
26. J. Benveniste, J.P. LeCouedic, J. Polonsky and M. Tence. Structural analysis of purified platelet activating factor by lipases, Nature, 269:170. (1977).
27. J.M. Marcum, M. McGill, E. Bastida, A. Ordinas and G.A. Jamieson. The interaction of platelets, tumor cells and vascular subendothelium. J. Lab. Clin. Med. 96:1046. (1980).
28. H.R. Baumgartner, The role of blood flow in platelet adhesion, fibrin deposition and formation of mural thrombi. Microvasc. Res. 5:167. (1973).
29. N.J. Weiss

30. V.T. Turitto, N.J. Weiss, H.R. Baumgartner, Platelet interaction with rabbit subendothelium in von Willebrand's disease, J. Clin. Invest. 74:1730-1741.
31. K.S. Sakariassen, J.D. Banga, P.G. Groot and J. Sixma. Comparison of platelet interaction with subendothelium of human renal and umbilical arteries and the extracellular matrix produced by human venous endothelial cells. Thromb. Haemostas. 52 (1) 60-65. (1984).
32. J.J. Sixma, Role of blood vessel, platelet coagulation interactions in haemostasis. In: Haemostasis and Thrombosis Bloom, A.L. Thomas D.P. (eds.) Churchill Livingstone, London, pp. 252-267.
33. V.T. Turitto and H.R. Baumgartner, Platelet-surface interactions, In: Haemostasis and Thrombosis., (ed) Coleman R.W. ed. J.B. Lippincott Company, p. 364-379. (1982).
34. D. Mohanty and P. Hilgard, A new platelet aggregating material (PAM) in an experimentally-induced rabbit fibrosarcoma. Thromb. Haemostas. 51:192. (1984).

FIBRINOLYSIS AND ITS EFFECT ON TUMOUR GROWTH AND METASTASIS

G.T.Layer, M. Pattison, B.Evans, D.R. Davies, and
K.G. Burnand

Departments of Surgery and Histopathology, St. Thomas's
Hospital Medical School, London, SE1 7EH

INTRODUCTION

Trousseau in the mid 19th century first observed the relationship between tumours and blood coagulation when he recognised
thrombophlebitis migrans as a manifestation of malignancy. Since
then the blood coagulation cascade and the fibrinolytic system have
been implicated in the growth and invasion of malignant tumours
and the shedding and implantation of metastases[1,2]. Some tumour
biopsies have been found to have a procoagulant activity[3] in that
they have clotted fresh plasma while others have been shown to have
a fibrinolytic activity when placed in contact with plasminogen-
rich fibrin [4,5,6]. However, the subject has been plagued with a
poor understanding of natural fibrinolysis, until the recent biochemical characterisation of plasminogen activator and the other
enzymes involved in the degradation of fibrin [7].

Fibrin has been seen accompanying tumour emboli in the rabbit
ear chamber[8] and recognised by its characteristic electron microscopic pattern around embolic malignant cells injected into the
rat tail vein [9]. In human tumours Dvorak[10] has identified a fibrin
network surrounding human breast carcinoma and has suggested that
this represents a host defence response. Fibrin has also been
seen surrounding small cell carcinomas of the lung [11].

Patients with certain tumours appear to benefit from long term
anticoagulation[12] and warfarin has doubled the two year survival
of many types of cancer[13]. Urokinase is reported to decrease the
recurrence rate of colorectal tumours following surgical excision[14].
Other fibrinolytic agents have prevented tumour growth in some
animal models[15] but in different tumour systems anti-fibrinolytics
have proved equally effective[16].

185

Interactions between the fibrinolytic activity in the blood, the vessel wall, interstitial fluid and tissues have been postulated to occur at other sites[17]. A fibrin barrier seen surrounding certain tumours may enhance invasion and metastasis by stabilising growth and implantation and preventing destruction by host defence systems[18]. The fibrinolytic capacity of tumour tissues may determine the ability of malignant cells to invade the host and in turn affect metastatic potential. An index of activation to inhibition of tumour cells has previously been reported to relate to the presence of metastases[19].

We have studied systemic and tissue fibrinolysis in the C57Black female mouse bearing the Lewis lung carcinoma; a well defined model of true spontaneous metastasis. This appears preferable to studying tissue explants injected as embolic "metastases" since these are not subjected to host influences which may effect cellular escape from an endogenous tumour. We have also studied systemic and tissue fibrinolysis in women with "early" carcinoma of the breast.

ANIMAL EXPERIMENTS

The effect of a drug known to enhance fibrinolysis in man was studied in mice before experiments could be devised to investigate the effect of altering sytemic fibrinolysis on the growth and spread of the tumour model.

Stanozolol, a synthetic anabolic steroid which enhances natural fibrinolysis in man[20] was administered by weekly depot intramuscular injection at 60 times the equivalent human dose (0.5mg/mouse/week), to the hind leg of 40 C57Bl female mice. The injection was made up in 0.1ml sterile water. Twenty mice received the drug and 20 controls received 0.1ml of sterile water. The mice were weighed and killed after five weeks by guillotine decapitation and the trunk draining blood (0.75 ml) was collected in a cold citrated wide-neck tube.

In subsequent experiments two different doses (2.5mg and 0.125mg) of stanozolol were given to two other groups of mice in order to determine whether a dose-response relationship existed.

The euglobulin clot lysis time (ECLT) and clot weight fibrinogen (CWF) were estimated as indicators of systemic fibrinolysis. The ECLT was measured by a modification of the technique described by Marsh[21] in which the ice-cold blood is acidified to pH5.9 and centrifuged. The precipitate is the euglobulin fraction containing fibrinogen, plasminogen and plasminogen activator and this is

dissolved in buffered saline and aliquots clotted with thrombin.
The spontaneous lysis time at 37°C expressed in minutes is known
as the ECLT. The clot weight fibrinogen[22] was estimated for
pooled blood from batches of five mice receiving the same treatment.

The liver, lungs and flank tissue were dissected from the body,
washed in ice-cold normal saline to remove excess blood, dried
carefully, then snap frozen using solid carbon dioxide and stored
at -70°C in Parafilm evelopes prior to the estimation of tissue
fibrinolysis. This was measured by our modification of the cryodestruction technique originally described by De Cossart and
Marcuson[23]. A standard weight of deep frozen tissue (0.05gm) is
taken and homogenised in a pre-cooled teflon cup containing a
tungsten ball on a Braun Mikrodismembrator. The cup is vibrated
for 30 seconds at 50 cycles per second then 2 ml of phosphate
beffered saline are added before a further 10 second vibration.
The resulting even suspension is used for analysis. Aliquots of
30 microlitres are placed in triplicate on plasminogen-rich fibrin
substrate plates at 37°C before the area of lysis is assessed by
measuring the maximum perpendicular diameters of the transparent
zones. The product is the representative area (A) and is related
to the weight (W) of the homogenate by the equation:

$$\log A = m \log W + \log k \quad \text{where m and k are constants} \quad [24]$$

Values at the standard weight of tissue are used in calculations.

RESULTS

The weight gain in both groups of mice was similar over the
five week period (41.6% drug : 38.5% control). The mice receiving
weekly stanozolol at all doses had a significantly prolonged ECLT
indicative of decreased blood fibrinolysis (tables 1a and 1b).
The CWF was significantly raised in animals receiving the standard
dose of stanozolol (0.5mg) but showed a variable response at the
other doses.

Tissue fibrinolytic activity could not be detected in the liver
or flank of either group of mice; but in the lungs where activity
was present, it was depressed by stanozolol. This reduction did
not reach statistical significance ($289mm^2$ drug, n=11 : $376mm^2$
control, n=18).

Having determined taht stanozolol predictably depresses
systemic fibrinolysis in the mouse, the effect of this alteration
was then investigated in tumour-bearing mice.

Table 1: The effect of stanozolol on systemic fibrinolysis in mice

a)

	CONTROLS water	STANOZOLOL 0.5mg	STANOZOLOL 2.5mg
Mean ECLT	18.8	29.5	160
Standard error	1.4	3.0	45.8
Significance compared to controls		p<0.001	p<<0.001
Mean pooled CWF	147	225	158
Standard error	7.9	9.0	5.1
Significance compared to controls		p=0.05	n.s.

b)

	CONTROLS water	STANOZOLOL 0.125mg	STANOZOLOL 0.5mg
Mean ECLT	18.2	28.9	37.0
Standard error	0.8	3.4	2.5
Significance compared to controls		p=0.009	p=0.001
Mean pooled CWF	117	147	222
Standard error	4.7	0.6	3.0

(number of pooled batches unsuitable for statistical analysis)
Mann Whitney U test throughout for statistical analysis

The Lewis lung carcinoma (3LL) was cultured and passaged in female C57B1 mice by inoculation of cell suspensions to the flank or by insertion of blocks of tissue subcutaneously. The tumour was obtained from the Institute for Cancer Research, Sutton, Surrey. Single cell suspensions were prepared by trypsinisation and addition of DNase[25] to fragments of fresh tumour.

A cell count was made in a haemocytometer chamber and 30 mice were injected subcutaneously in the flank with 1.3×10^5 cells. Ten of the 30 mice in the experiment had received stanozolol prior to tumour inoculation, and these together with ten others received stanozolol at weekly intervals after tumour inoculation. Ten other mice received water and acted as controls. Ten additional mice did not receive tumour and were divided into two groups, half receiving stanozolol, and half receiving water. All the mice were fed on a small animal diet with free access to water and they were weighed weekly. The growing tumours were measured by their two perpendicular maximum diameters and the weight calculated by the standard formula of the National Cancer Institute:

$$\text{weight} = \text{length} \times (\text{width}^2) / 2 \ [26].$$

Blood, tissues and tumour were collected at five weeks when the animals were killed by decapitation. The visible pulmonary metastases were counted by insufflating the intubated lungs with a low pressure air supply[27]. The metastasis count in this experimental model is recognised as a measure of the primary tumour's ability to metastasise, despite the variability observed.

RESULTS

Systemic fibrinolysis in the animals in the control and the stanozolol treated groups indicated a quenching effect by the presence of tumour[28] with decreased fibrinolysis and raised fibrinogen (table 2).

The ECLT in mice receiving stanozolol alone was again signif- icantly prolonged compared with controls, although both times were faster than in the previous experiment. This may have been due to less satisfactory clot precipitation than was obtained in the initial studies (table 1).

The tumour tissue itself displayed only minimal tissue fibrinolysis. The central portion of these growths was very necrotic and dead cells do not display any fibrinolytic activity.

Three mice in each group bearing tumour died at similar times from haemoperitoneum following rupture of a tumour which had clearly been implanted too deeply. Although there was an increase in primary tumour weight and number of metastases in the group treated with stanozolol after tumour inoculation, this difference did not reach significance when compared with the control group. However, the group of mice that received stanozolol prior to tumour inoculation displayed both a significant increase in the primary tumour weight and in the number of pulmonary metastases when compared with controls (table 3).

Table 2: The effect of tumour on systemic fibrinolysis

	WATER ALONE	STANOZ- OLOL ALONE	TUMOUR BEARING CONTROLS	STANOZ- OLOL PRE/POST TUMOUR	STANOZ- OLOL POST TUMOUR
Mean ECLT	10.0	15.4	63.1	52.0	37.2
Standard error	1.1	1.4	22.9	25.0	13.4
Significance					
-compared to water alone	$p<0.05$		p=0.009	$p<0.02$	p=0.008
-compared to stanozolol alone			n.s.	n.s.	n.s.
No significance detected between ECLT of tumour bearing groups					
Mean CWF	32	119	135	142	123
(number of pooled batches unsuitable for statistical analysis)					

Table 3: Primary tumour size and number of metastases in
stanozolol treated and untreated controls

7 mice per group	TUMOUR BEARING CONTROLS	STANOZOLOL PRE AND POST TUMOUR	STANOZOLOL POST TUMOUR
Mean primary weight	2986	5037	3475
Standard error	661	806	1110
Significance compared to controls		p<0.03	n.s.
Mean no. metastases	12.9	28.4	23.7
Standard error	4.8	5.1	4.3
Significance compared to controls		p<0.02	n.s.

DISCUSSION

We were surprised to find that stanozolol prolonged the ECLT
(the opposite to man) but further investigations have confirmed
that the mouse fibrinolytic system differs from man in its response
to fibrinolytic manipulation. This has recently been confirmed
by others[29].

Stanozolol promotes growth of the primary tumour and increases
the number of metastases formed, but interestingly there was no
correlation between the weight of the primary tumour and the number
of metastases. Stanozolol may have affected primary growth and the
shedding of metastases by different mechanisms.

A direct effect of the drug other than that by decreasing
fibrinolysis in the pre-treated animals has not been excluded but
an anabolic effect of stanozolol is unlikely as weight gain was
not found in the treated animals.

EXPERIMENTS WITH THE BREAST CARCINOMA

Fresh tumour tissue was obtained for this study from 18
patients undergoing mastectomy and axillary clearance for carcinoma
of the breast. Tissue fibrinolytic activity was estimated by the
cryodestruction technique described above for samples from the tumour
centre, the tumour-host interface and the "normal" surrounding
epithelium. The results were related to the tumour size and the
presence of regional node metastases. Two samples were taken
from the tumour centre, the host-tumour interface and the surrounding
breast. Tissue was also obtained from eight breasts with benign
disease.

Table 4: Fibrinolytic activity of malignant and benign breast tissues

	BENIGN CONTROLS	"NORMAL" SURROUND	TUMOUR INTERFACE	TUMOUR CENTRE
Mean fibrin plate lysis area mm^2	485	579	646	459
Standard error	44	57	64	54
Significance compared to centre	p<0.004	p<0.003	p<0.011	

RESULTS

The mean clot weight fibrinogen of the 18 patients with breast cancer taken pre-operatively was raised above the normal range but there was no abnormality of the euglobulin clot lysis time. Per-operative fibrinolysis was increased as previously described[30] and this returned to pre-operative levels within a week.

The tumour-host interface and the surrounding breast were found to have a significantly higher level of fibrinolytic capacity compared to benign controls and these all had significantly greater activity than the tumour centre (table 4).

Haematoxylin and eosin stains of the tumours confirmed invasive scirrhous ductal cancer with no evidence of central necrosis or host infiltration to account for the activity difference at the tumour centre. Preliminary immunohistochemical mapping studies using polyclonal antibodies to the fibrinolytic enzymes, tissue plasminogen activator (tPA) and urokinase (UK), have shown localisation of tPA-antibodies to tumour cells with minimal fixation to stromal elements. Cells surrounding clumps of tumour cells stained differentially with UK-antibodies but not with tPA-antibodies.

DISCUSSION

In this study we have shown that malignant breast tumours do not necessarily have a higher activity than benign or "normal" surrounding breast tissues but we have demonstrated a rise in fibrinolytic activity from the centre to the tumour-host interface. This rise may be related to fibrin distribution or interactions in growth and invasion. Subsequent evaluation of the data suggests that high levels of tissue activity are found in tumours without evidence of local lymphatic spread, and conversely, low levels are associated with node involvement

Our studies in mouse and man provide evidence for the involvement of the fibrinolytic system in tumour growth and spread and suggest that alterations in the host environment by systemic agents may influence tumour growth. The prospect of disturbing the balance between host and tumour tissue fibrinolytic activity by systemic agents to alter the growth characteristics of a malignant tumour is attractive and further experiments in the mouse tumour model using drugs which enhance mouse systemic fibrinolysis are in progress. It is conceivable that adjuvant fibrinolytic enhancement therapy combined with surgery may be of benefit in preventing or delaying tumour spread.

REFERENCES

1.   M.B. Donati, A. Poggi and N. Semeraro, Coagulation and malignancy, in: "Recent Advances in Blood Coagulation 3," Leon Poller, ed., Churchill Livingstone, Edinburgh (1981).
2.   H. Rasche and M. Dietrich, Haemostatic abnormalities associated with malignant diseases, Eur.J.Cancer 13:1053 (1977).
3.   R.A.Q. O' Meara, Coagulative properties of cancers, Ir.J.Med.Sci. 396:474 (1958).
4.   E.E.Cliffton and C.E. Grossi, Fibrinolytic activity of human tumours as measured by the fibrin plate method, Cancer 8:1146 (1955).
5.   B. Nagy, J. Ban and B. Brdar, Fibrinolysis associated with human neoplasia: production of plasminogen activator by human tumours, Int.J.Canc. 19:614 (1977).
6.   J.D. Franklin, A.S. Gervin, D.G. Bowers and J.B. Lynch, Fibrinolytic activator activity in human neoplasms, Plast. Reconst.Surg. 61(2):241 (1978).
7.   S.A. Cederholm-Williams, S. Houlbrook, N.W. Porter, J.M. Marshall and H. Chissic, Characterisation of plasminogen activators secreted by human malignant cells, in: "Treatment of Metastasis: Problems and Prospects", K. Hellman and S.A. Eccles, eds., Taylor and Francis, London (1985).
8.   J.H. Johnson and S. Wood Jr., An in vivo study of fibrinolytic agents on V2 carcinoma cells and intravascular thrombi in rabbits, Bull.John Hopk.Hosp. 113:335 (1963).
9.   E.C. Chew and A.C. Wallace, Demonstration of fibrin in early stages of experimental metastases, Canc.Res 36:1904 (1976).
10.   H.F.Dvorak, G.R. Dickersin, A.M. Dvorak, E.J. Manseau and K. Pyne, Human breast carcinoma: fibrin deposits and desmoplasia, J.Natl.Canc.Inst 67(2):335 (1981).
11.   L.R. Zacharski, A.R. Schned and G.D. Sorenson, Occurence of fibrin and tissue factor antigen in human small cell carcinoma of the lung, Canc.Res 43:3963 (1983).

13. P. Hilgard and R.D. Thornes, Anticoagulants in the treatment of cancer, Eur.J.Canc. 12:755 (1976).

14. H. White, J.D. Griffiths and A.J. Salsbury, Circulating malignant cells and fibrinolysis during resection of colrectal cancer, Proc.Roy.Soc.Med. 69:467 (1976).

15. P. Hilgard, H. Schulte, G. Wetzig, G. Schmitt and C.G. Schmidt, Oral anticoagulation in the treatment of a spontaneously metastasising murine tumour (3LL), Br.J.Canc. 35:78 (1977).

16. H. I. Peterson, Fibrinolysis and antifibrinolytic drugs in the growth and spread of tumours, Canc.Treat.Rev. 4:213 (1977).

17. N.L. Browse, L. Gray, P.E.M. Jarrett and M. Morland, Blood and vein-wall fibrinolytic activity in health and vascular disease, Br.Med.J. 1:478 (1977).

18. H.F. Dvorak, D.R. Senger and A.M. Dvorak, Fibrin as a component of the tumour stroma: origins and biological significance, Canc.Met.Rev. 2:41 (1983).

19. J.M. Malone, A.S. Gervin, W.S. Moore and K. Keown, Tumour interaction with the fibrinolytic system, J.Surg.Res. 26:581 (1979).

20. K.G. Burnand, G. Clemenson, M. Morland, P.E.M. Jarrett and N.L. Browse, Venous lipodermatosclerosis: treatment by fibrinolytic enhancement and elastic compression, Br.Med.J. 28:7 (1980).

21. G.I.C. Ingram, The determination of plasma fibrinogen by the clot weight Method, J.Biochem. 51:583 (1952).

22. N. Marsh, "Fibrinolysis", John Wiley and Sons, Chichester, New York, Brisbane and Toronto (1981).

23. L. De Cossart and R.W. Marcuson, A simple quantitative assay of tissue plasminogen activator, J.Clin.Path. 35:980 (1982).

24. T. Astrup and J. Jespersen, in: "Progress in Fibrinolysis VI", J.F. Davidson, F. Bachmann, C.A. Bouvier and E.K.O. Kruithof, eds, Churchill Livingstone, Edinburgh (1983).

25. T.C. Stephens, K. Adams and J.H. Peacock, Metastasis of Lewis Lung Carcinoma regrowing after cytotoxic treatments, in: "Proceedings of the E.O.R.T.C. Conference on Clinical and Experimental Aspects of Metastasis", Martinus Nighoff, The Hague (1980).

26. National Cancer Institute, Protocols for screening chemical agents and natural products against animal tumours and other biological systems, Canc.Chemo.Rep.III 3(2):1 (1972).

27. G.T. Layer, M. Pattison and K.G. Burnand, Does fibrinolysis have a role in tumour growth and spread?, in: "Treatment of Metastasis: Problems and Prospects", K. Hellman and S.A. Eccles, eds, Taylor and Francis, London (1985).

28. J.A.N. Rennie and D. Ogston. Fibrinolytic activity in malignant disease, J.Clin.Path. 28:872 (1975).

29. Beecham Research Laboratories, personal communication (1984).

30. N.L. Browse, L. Gray and M. Morland, Changes in the blood fibrinolytic activity after surgery (effect of deep vein thrombosis and malignant disease), Br.J.Surg. 64:23 (1977).

PERICELLULAR PROTEOGLYCANS AND METASTASIS FORMATIONS: MODULATION

BY HYDROCORTISONE AND GROWTH FACTORS

E. Moczar, M. Becker*, and M.F. Poupon*

Laboratoire de Biochimie du Tissu Conjonctif, CNRS
GR.40, 8 rue du Général Sarrail, 94010 Créteil & *IRSC
du CNRS, ER 278, BP 8, F-94802 Villejuif, France

INTRODUCTION

To form metastases, tumor cell have to free themselves from the surrounding tumoral tissue and pass through biological barriers, such as the endothelial cell layers and its underlying basal lamina[1,2,3]. Before the enzymatic destruction of the tissue barriers, malignant cells had to recognize the vascular endothelial cells and the subendothelial matrix components, and adhere to them[4,5,6,7]. It is generally accepted that the carbohydrate moieties of proteoglycans and glycoproteins are involved in the regulation of cell recognition, adhesion and differentiation. Alteration of the carbohydrate moieties of these macromolecules, on the cell surface affect developpement of neoplasms, modifying the interaction of the cells with their environment.

We have noticed that hydrocortisone (HC) induced an advanced stage of differentiation when added to the culture medium of tumor cells, originated from a nickel-induced rhabdomyosarcoma (RMS) of rat. Parallely, HC given to rats bearing the same RMS, in the course of tumor growth and after its surgical ablation, prevented the metastatic invasion of lungs and lymph nodes[8,9]. On the other hand, epithelial and eye-derived growth factors (EGF and EDGF respectively) retard the deregulation of the collagen and the proteoglycan biosynthesis and cell dedifferentiation[10]. It was found that rat rhabdomyosarcoma sublines with different metastatic potentials are stimulated in a different way by these two growth factors[9].

Considering these findings, it appeared reasonable to ask,

195

whether cell lines of different metastatic potentials exhibit
differences in the sugar moieties of cell surface proteoglycans
and whether the formation of the glycosaminoglycan side chains of
these proteoglycans can be modulated by hydrocortisone and by
growth factors.

MATERIAL AND METHODS

A rhabdomyosarcoma (RMS) has been induced by I.M. injection
of Nickel in a Wistar AG rat as previously described[8]. The
primary tumor has been excised and set up in culture. A parental
cell line named 9-4/0 has been derived, and subsequently cloned as
reported[11]. Three sublines were used here, line 9-4/0 a highly
metastazing one, subline 8, a weakly metastasizing and weakly
colonizing clone and line 13 not metastasizing but colonizing one.
All the lines were cultivated in Dulbecco's modified medium DMEM
supplemented with 10% fetal calf serum (FCS) and antibiotics.
Characteristics of the lines were defined in a constant range of
in vitro passages. HC was added to the culture medium in 0.5 ug/ml
final concentration. The cell lines 9-4/0 and 13 were stimulated
with FGF (Fibroblast growth factor) and the line 8 with EDGF (Eye
derived growth factor) as described[9].

Cells were metabolically labelled with (^{3}H)-glucosamine and
(^{35}S)sulfate for 5 days, changing the culture medium on the 3rd
day. The medium (extracellular compartment) was decanted when
cells reached confluency. The cell surface components were
successively extracted by heparin (100 ug/ml) which took of
pericellular membrane components and then by trypsin (2.5 ug/ml)
which enzymatically detached macromolecules more deeply
anchored[12]. The cellular residue was detached and solubilized by
triton. The culture medium, the heparin extract, the trypsin
extract and the cellular residue were analysed separately. The
glycosaminoglycan (GAG) chains of the proteoglycans were assayed
as described[11]. The glycoproteins and the proteoglycans from an
aliquot were precipitated together with trichloroacetic acid
(TCA). An other aliquot was digested by pronase, and precipitated
by cetylpyridinium chloride (CPC). Successive digestion of pronase
digest by chondroitinase ABC and nitrous acid, allowed to evaluate
the relative quantities of heparan sultate, chondroitin sulfate
and hyaluronic acid in the GAG fraction. The radioactivity in the
precipitates was determined by scintillation counting. The
substraction of the radioactive value corresponding to the CPC
precipitate (proteoglycan) from the TCA precipitate allowed an
evaluation of the glycoprotein content. This difference between
the TCA precipitable and CPC precipitable labels represents the
glucosamine incorporated in the different glycoproteins
synthetized by the cell : basement membrane collagens,
fibronectine laminine and other intra and extracellular
glycoproteins.

RESULTS AND DISCUSSION

The main differences between the distribution of the labels in the glycoprotein and in the glycosaminoglycans of the pericellular fractions of the three cell lines studied are shown on the table 1. The results presented indicate, that the three lines originated from the same tumor, differed in the expression of cell surface components. The strongly metastatic and colonizing 9-4/0 line was characterized by a low content of cell surface glycoproteins and glycosaminoglycans related to the two very weakly metastatic (8 and 13) sublines.

Table 1. Effect of hydrocortisone and of growth factors on the incorporation of the (^{3}H) glucosamine in the glycoproteins and glycosaminoglycans of the two pericellular fractions, the heparin and the trypsin extracts of the rat rhabdomyosarcoma sublines.

	Line 9-4/0 GP CS HS HA	Subline 8 GP CS HS HA	Subline 13 GP CS HS HA
	HEPARIN EXTRACT		
C	150 120 60 140	600 300 200 150	330 65 100 50
HC	640 110 85 130	480 200 166 200	330 65 100 50
GF	170 90 100 140	270 70 110 130	300 290 200 200
	TRYPSIN EXTRACT		
C	300 50 102 51	570 420 270 50	170 180 220 40
HC	320 360 250 72	510 450 200 40	200 180 140 10
GF	400 208 110 41	960 218 160 81	228 91 129 30

Results are expressed in dpm x 10^{-3} x 10^{-5} cell. and are means of determinations from 4 culture dishes. The standard deviation was alway less then 10%. CS = chondroitin sulfate, HS = heparan sulfate, HA = hyaluronic acid. GP = glycoproteins.
C = control, HC = hydrocortisone-treated, GF = growth factor treated cells.

Immunofluorescence experiments demonstrated the decrease of fibronectine in the presence of HC in the 9-4/0 line. Thus the increased glycoprotein levels should be attributed to a glycoprotein different from fibronectin or a change in the accessibility of the antigenic determinants of the fibronectin. The surface of the cells from the very weakly metastatic and weakly colonizing 8 subline showed a higher level of glycoproteins and glycosaminoglycans in both pericellular fractions. The weakly metastatic, but strongly colonizing 13 subline showed also higher levels of radioactivity in the glycoproteins and glycosaminoglycans but only in the trypsin extract.

Hydrocortisone treatment induced significant changes in the 9-4/0 cells, while changes are generally not significant in the sublines 8 and 13. The distribution of the labels in the macromolecules studied in the HC-treated 9-4/0 cells approches that found in the non-treated 8 subline.

The growth factors used increased the label in the cell surface glycosaminoglycans of the 9-4/0 line, but decreased in the 8 subline. In the 13 subline, the growth factor increased the label in the sulfated glycosaminoglycans in the heparin extract but decreased in the trypsin extract.

The variation of the distribution of label in the macromolecular components studied was less accentuated in the extra- and intracellular compartments (data not shown) then in the pericellular fractions.

The increased label in the glycosaminoglycans of the not metastatic but colonizing 13 subline, suggests that the cell-surface proteoglycans are also implicated in the first step of the metastasis formation ; in the escape of the cells from the original tumor. It has been found that the cells of the very weakly invading 8 subline adhere less rapidly to a monolayer of vascular endothelial cells as the highly metastatic 9-4/0 parental lines[9]. It was shown on the other hand, that the accumulation of chondroitin sulfate proteoglycans and of hyaluronic acid on the cell surface is an important step in the detachment of the cells from various substrates[13]. Thus, it is tempting to speculate, that cell surface proteoglycans are involved in the mechanism of the attachment and of the recognition between the vascular endothelium and the metastatic tumor cells. It is possible, that beside their direct participation in cell-cell and cell-matrix interactions, glycosaminoglycans are implicated also by other ways in the control of metastasis formation, p.e. by inhibiting proteolytic or heparan sulfate degrading enzymes or by modulating the regulation of intracellular events[14,15].

It was shown recently that the biosynthesis of sulfated glycosaminoglycans decreases in highly metastatic lymphoma cell lines[16] and that of the chondroitin sulfate in a highly metastatic

Lewis-lung tumor cell line (unpublished results).

The results presented suggest that the accumulation of the proteoglycans, especially that of the chondroitin sulfate proteoglycans on the cell surface, induced by the hydrocortisone, may be involved in the decrease of the metastatic potential of the rhabdomyosarcoma cells. Further studies are however necessary to elucidate the nature of glycoproteins and of the proteoglycans synthetized by the different cell lines and their role in the tumor cell invasion and metastasis.

ACKNOWLDEGEMENTS

This work was supported by the CNRS (Gr. N° 40 and ER 278) and the Association du Developpement de la Recherche sur le Cancer, Grant N° 1225 and 6343 to M.-F. Poupon and E. Moczar respectively.

REFERENCES

1.   R.H. Kramer, K.G. Vogel and G.L. Nicolson, Solubilization and degradation of subendothelial matrix glycoproteins and proteoglycans by metastatic tumor cells, J. Biol. Chem. 257:2678 (1982).

2.   G.L. Nicolson, Organ colonization and the cell-surface properties of malignant cells, Bioch. Biophys. Acta 695:113 (1982).

3.   I. Vlodavsky, Y. Ariav, R. Atzmon, Z. Fuks, Tumor cell attachment to the vascular endothelium and subsequent degradation of the subendothelial extracellular matrix, Exp. Cell. Res. 140:149 (1982).

4.   J. Montreuil, Primary structure of glycoprotein glycans, basis for the molecular biology of glycoproteins, in "Advances in carbohydrate chemistry and biochemistry", Academic Press, London 37:157 (1980).

5.   R. Iozzo, Proteoglycans and neoplastic-mesenchymal cell interactions, Human Pathol. 15:2 (1984).

6.   G. Greenburg and D. Gospodarowicz, Inactivation of a basement membrane component responsible for cell proliferation but not for cell attachment, Exp. Cell. Res. 140:1 (1982).

7.   E. Moczar, J. Tassin and Y. Courtois, Interaction of bovine epithelial lens (BEL) cells with extracellular matrix (ECM) and eye-derived growth factor (EDGF) III. Control of glycoprotein and proteoglycan synthesis, Exp. Cell. Res. 149:95 (1983).

8.   M. Becker, E. Moczar, V. Lascaux and M.F. Poupon,
      Relationship between in vitro effects and in vivo
      control of metastasis induced by "hydrocortisone in a
      rat rhabdomyosarcoma model", _in_ "Treatment of
      Metastasis : Problems and Prospects" K. Hellman and
      S.A. Eccles eds Taylor and Francis, London and
      Philadelphia p. 16 (1984).

9.   M.F. Poupon, M. Becker, C. Pauwels, E. Moczar and S.
      Korach, Etude expérimentale des métastases des
      cancers. Bull. Cancer (Paris) 71:453 (1984).

10.   J. Tassin, E. Jacquemin and Y. Courtois, The
      interaction of extracellular matrix and EDGF with
      bovine lens epithelial cells. Effects on short time
      adhesiveness and long term organization in culture,
      Exp. Cell. Res. 149:69 (1983).

11.   F.L. Sweeney, J. Pot-Deprun, M.F. Poupon and I.
      Chouroulinkov, Heterogeneity of the growth and
      metastatic behavior of cloned cell lines derived from
      a primary rhabdomyosarcoma, Cancer Res. 42:3776 (1982).

12.   L. Kjellen, A. Oldberg and M. Hook, Cell-surface
      heparan sulfate, J. Biol. Chem. 255:10407 (1980).

13.   B.J. Rollins and L.A. Culp, Glycosaminoglycans in the
      substrate adhesion sites of normal and virus-
      transformed murine cells, Biochemistry 18:141
      (1979).

14.   K. Furukawa and V.P. Bhavanandan, Influence of
      glycosaminoglycans on endogenous DNA synthesis in
      isolated normal and cancer cell nuclei, Biochem.
      Biophys. Acta 697:344 (1982).

15.   V. Schirrmacher and 15 authors, Importance of cell
      surface carbohydrates in cancer cell adhesion,
      invasion and metastasis, Invasion and Metastasis
      2:313 (1982).

16.   R.T. Schwartz and V. Schirrmacher, Different
      expression and shedding of glycosaminoglycans by high
      and low metastatic lymphoma cells, Proc. of Bat-Sheva
      Seminar on Tumor Metastasis : Control Mechanisms, May
      8-13, p. 63. The Weizmann Institute of Sciences,
      Rehovot, Israel (1983).

SECTION 5

CELLULAR INTERACTIONS: GRANULOMAS, HEALING WOUNDS,

DRESSINGS AND SUTURES

THE MONONUCLEAR PHAGOCYTE SYSTEM IN GRANULOMAS

John L. Turk

Department of Pathology
Institute of Basic Medical Sciences
London WC2A 3PN

SUMMARY

A comparison is made between immunologically induced and non-immunologically induced granulomas in guinea pigs injected with metals (zirconium and aluminium) or with mycobacteria (BCG vaccine and Mycobacterium leprae). Immunological granulomas were characterised by epithelioid cells and fibrosis, whereas non-immunological granulomas contained phagocytosing macrophages with little evidence of fibroblast activation. Epithelioid cells carry the same specific macrophage antigen as phagocytosing macrophages and this can be detected by the use of a specific monoclonal antibody. However, they differ from phagocytosing macrophages in that they are poorly phagocytic, not glass adherent and lack Ia antigen. They are, however, secretory cells with rouch endoplasmic reticulum. A relation between the presence of these cells and increased collagen synthesis is indicated.

A study of accessory cell function showed that the epithelioid cells of BCG granulomas were unable to support mitogen but not antigen induced proliferation of T-lyphocytes. The macrophages of M. leprae granulomas did not support either a mitogen or antigen induced proliferative response.

INTRODUCTION

The term granuloma was introduced by Virchow [1] to describe certain well circumscribed swellings consisting of 'granulation tissue' found in a number of chronic infectious diseases including tuberculosis, leprosy, syphilis and leishmaniasis. Later authors [2] distinguished large swollen ipithelial-like cells that they called

203

epithelioid cells from the lymphoid elements in these lesions. Metchnikoff [3] formally recognised that these cells were related to the new type of cell that he called - the macrophage. The relation between these granulomatous diseases and the presence of delayed hypersensitivity was shown by Zinsser [4]. Cells of the mononuclear phagocyte series are now considered to be the most important feature of all granulomas. However, it is also now accepted that granulomas need not necessarily be associated with infectious diseases or with a state of delayed hypersensitivity. In addition to infectious diseases hypersensitivity granulomas can be produced by the metals beryllium and zirconium and are also found in sarcoidosis, a disease in which the role of an infectious agent has yet to be identified. Non-immunological granulomas occur in silicosis and following the injection of metallic compounds such as aluminium hydroxide and kaolin. However, they can also occur in certain states of chronic infections when there is a specific failure of cell-mediated immunity. Examples of these are lepromatous leprosy and the systemic form of leishmaniasis (Kala-azar) and the systemic mycoses. The granulomas of tuberculoid leprosy and localised forms of leishmaniasis are examples, however, of immunological granulomas and are associated with strong delayed hypersensitivity reactions. The predominant cell that features in non-immunological granulomas is the typical phagocytosing macrophage. In lepromatous leprosy, this foamy cell often referred to as the 'Virchow cell' is packed full of 'globi' of M. leprae. In immunological granulomas the 'epithelioid cells' are, however, poorly phagocytic. In addition to epithelioid cells, hypersensitivity granulomas contain lymphocytes, other inflammatory cells and fibroblasts leading to intense fibrosis. The relation between epithelioid cells and phagocytosing macrophages is one of the more important aspects of hypersensitivity of granulomas. It is important to ask the question why epithelioid cells rather than phagocytosing macrophages are a particular feature of hypersensitivity granulomas. In addition, the function of the epithelioid cells and its relation to fibroblast activity forms a further important field of study. The present communication will attempt to answer some of these points in studies using experimentally induced granulomas in the guinea pig.

EXPERIMENTAL MODELS OF HYPERSENSITIVITY GRANULOMAS

Metal Granulomas

Although it is possible to induce non-immunological granulomas by the injection of aluminium compounds directly into the skin, the injection of similar zirconium compounds does not have this effect. However, if guinea pigs are sensitized with sodium zirconium lactate in Freund's adjuvant, they first develop delayed hypersensitivity reactions. After a number of weeks of skin testing, these delayed hypersensitivity reactions may develop into nodular granulomas containing giant cells and epithelioid cells, as well as other inflammatory cells including lymphocytes. Thus, there appears in this situation to be a direct relation between delayed hypersensitivity and granuloma formation induced by this particular compound [5]. In the case of the aluminium hydroxide granulomas, there is no evidence of lymphycytic infiltration and the cells of the mononuclear phagocyte series are actively phagocytosing macrophages. Using electron microscopy, many of the cells of the mononuclear phagocyte series in what is a typical epithelioid cell granuloma associated with delayed hypersensitivity to zirconium have a distinct appearance. Apart from a fimbriated cell periphery and interdigitations with similar cells, these cells have a marked rough endoplasmic reticulum and golgi apparatus [6]. This would indicate a predominantly secretory activity. Another strong difference between the zirconium hypersensitivity granuloma and the non-immunological granuloma induced by aluminium hydroxide is that the hypersensitivity lesion is characterised by the presence of large numbers of fibroblasts and rapid resolution. The non-immunological lesion induced by aluminium, showed poor resolution and persisted for many months. This would indicate the possibility that secretory epithelioid cell formation was associated in some way with fibroblast activation.

Experimental Mycobacterial Granulomas

The granuloma of tuberculoid leprosy is a typical epithelioid cell granuloma associated with the presence of a strong state of delayed hypersensitivity, and the ability to develop epithelioid cell granulomas (Mitsuda reaction) when the individual is injected intradermally with heat killed M. leprae. This indicates that this granuloma is also produced by a cell-mediated immune response. In contrast the granuloma of lepromatous leprosy consists solely of macrophages that have ingested 'globi' of M. leprae and there is an associated specific defect in cell-mediated immunity and patients are unable to develop an epithelioid cell granuloma when injected intradermally with the Mitsuda reagent.

Models of these two types of granuloma have been developed in the auricular lymph node of guinea pigs injected in the dorsum of the ear with Mycobacteria [7]. The intradermal injection of BCG vaccine, live or Cobalt irradiated, produced typical epithelioid cell granulomas in which the epithelioid cells were shown by electron microscopy to contain rough endoplasmic reticulum. These cells were similar to those seen in the zirconium granulomas. There were also considerable numbers of fibroblasts and evidence of new collagen formation. In contrast, the granulomas produced by the injection of Cobalt irradiated M.lepra consisted mainly of macrophages which still contained ingested M.leprae. There was little evidence of phagocytosed bacteria in the cells of the mononuclear phagocyte series in the BCG granulomas. Quantitative evaluation of these granulomas according to the weight of the lymph nodes and the area of granulomatous infiltration, as measured by planimetry, indicated

that the BCG granulomas reached their peak two weeks after induction,
whereas the M.leprae granulomas peaked at 5 weeks.

Comparison was made of a number of parameters, between cells of
the mononuclear phagocyte series in these two types of granuloma and
oil induced peritoneal macrophages (Table 1). Both peritoneal macrophages and M.leprae macrophages were glass adherent. However, the
epithelioid cells of the BCG granulomas did not adhere to glass.
This was consistant with their failure to phagocytose. EA and EAC
rosetting on the peritoneal cells showed that a large proportion of
these macrophages carried surface receptors for the Fc component of
IgG and C3. A high percentage of these macrophages also exhibited
peroxidase and non specific esterase activity. Immunofluorescence
using FITC conjugated monoclonal antibody against human fibronectin
in these macrophages. A high percentage of cells of the MPS infiltrating the BCG and M.leprae induced granulomas were esterase positive
and also showed the presence of fibronectin. However, these cells
did not carry Fc or C3 surface receptors nor did they exhibit peroxidase activity.

Despite the general acceptance that epithelioid cells were
related to mononuclear phagocytes [3], it was important to demonstrate
a formal immunological relationship between these two cell types, as
the term epithelioid cell had been used originally to cover all non-
lymphoid mononuclear cells in granulomas whether of an immunological
or non-immunological nature. A monoclonal anti-guinea pig macrophage
antibody was, therefore, prepared using guinea pig peritoneal macrophages as antigen. This was specific for macrophages and not just
directed against the Ia antigen or the Fc receptor, as it did not
stain LC2 leukemia cells (a B-cell line with Fc receptors and Ia
antigens) and did not block EA rosetting. This was compared with
an anti guinea pig Ia monoclonal antibody kindly given by Dr. Ethan
Shevach of the NIAID, NIH, Bethesda, Md. in immunofluorescence and
immunoperoxidase studies [8]. Both BCG epithelioid cells and M.leprae
macrophages stained with this reagent as did peritoneal exudate
macrophages and Kupffer cells, Langerhans cells and the M.leprae
granuloma macrophages were Ia positive. However, epithelioid cells
were Ia negative. Thus, it would appear that epithelioid cells are
related antigenically to other cells of the mononuclear phagocyte
series, including peritoneal and other macrophages. They differ in
that they are poorly phagocytic and thus not glass adherent, and lack
Ia antigen, indicating that they do not possess an antigen presenting
function. The presence of a rough endoplasmic reticulum would probably indicate that these cells have a secretory rather than phagocytic
function.

In order to see whether there was a relationship between epithelioid cell formation and increased fibroblast activity, collagen
synthesis was examined in explants of auricular lymph nodes from
BCG and M.leprae injected animals and compared with that in auricular

lymph nodes from animals painted on the dorsum of the ear with 2.4 dinitrofluorobenzene. Lymph nodes were cut into small pieces and incubated with ^{14}C-proline for 24 hours at 37°. They were then homogenised in tris-buffered water and divided into two aliquots. One was solubilised directly in 10% TCA for counting the total protein (T). The other was treated with collagenase at 34° for 90 minutes and the TCA precipitate solubilised to give a hydrolysed fraction (H) that did not contain collagen. The total protein less the hydrolysed fraction would then give the ^{14}C-incorporated into collagen. In these studies it was evident that the nodes from animals injected with BCG synthesise high levels of collagen as compared with the nodes from animals sensitized with DNFB or injected with Cobalt-irradiated M.leprae P. There was, therefore, a direct association between the presence of secretory epithelioid cells and the subsequent development of fibrosis. One possible explanation for the association between epithelioid cell granulomas and fibrosis is that fibroblast activation results from a release of a specific activating factor from the epithelioid cells with rough endoplasmic reticulum. Further experiments were therefore performed to see whether such a factor was released by granuloma tissues in culture.

Table 1. PROPERTIES OF LARGE CELLS INFILTRATING GRANULOMAS AS COMPARED WITH PERITONEAL EXUDATE MACROPHAGES

	Peritoneal exudate cells 72 hr oil induced	BCG (epith-elioid cells (2wk granuloma)	M. leprae macrophages (5wk granuloma)
Glass or plastic adherence	++ (c.60%)	± (c.10%)	+++ (c.70%)
Fc receptors	+++ (83%)	−	± (c.10%)
C3 receptors	+++ (73%)	−	± (c.10%)
Peroxidase	+++	−	−
Non-specific esterase	+++	++	++
Fibronectin	++	++	++
Specific macro-phage antigen	++ (75%)	+++ (100%)	++ (80%)
Ia antigen	+ (35%)	− (0%)	++ (80%)

Supernatants from both BCG and M.leprae granulomas were found to release soluble non-dialysable factors in vitro, which stimulated ^{14}C-leucine incorporation in fibroblasts in culture and depressed their ^{3}H-thymidine uptake. These supernatants did not show any

detectable macrophage migration inhibitory activity in vitro. On
the other hand, supernatants from sensitized lymphycytes incubated
with tuberculin-PPD had no effect on fibroblasts. Supernatants from
DNFB sensitized lymph nodes also showed stimulation of ^{14}C-proline
incorporation into total protein synthesised by fibroblasts and
depressed ^{3}H-thymidine incorporation. It would appear, therefore,
that fibroblast activation in lymph nodes containing mycobacterial
granulomas could result from the release of soluble factors of lymphocyte origin rather than from cells of the mononuclear phagocyte
series. These factors appear to be independent of classical lympho-
kines that act on macrophages in vitro The identification of these
factors has, however, not clarified the mechanism of fibroblast
activation in BCG granulomas, as compared with M.leprae granulomas,
nor has it added to our knowledge of the function of the secretory
epithelioid cell [10].

    Despite this, it is clear that the epithelioid cells form a
distinct subpopulation of cells of the mononuclear phagocyte function
for a secretory role. They are, therefore, not glass adherent and
lack Ia antigens. As they can be recognised in tissues mainly by
their presence cannot be determined by light microscopy alone. The
role these cells play in the formation of certain granulomas is
intriguing and required further study.

The Accessory Cell Function of Granuloma Cells

    In view of the fact that the epithelioid cells of BCG granulomas
were found to be mainly Ia antigen positive and the macrophages of
M.leprae granulomas were Ia negative, it was important to see whether
these cells could act as "accessory cells". An "accessory cell" may
be defined as a cell of the mononuclear phagocyte series or a dendritic
cell that plays a role in antigen presentation and whose presence is
necessary for T-lymphocyte proliferation. In this study it was possible to show that the total epithelioid cell population of BCG granulomas were able to support a mitogen (Concanavalin A) induced proliferative response of macrophage depleted autologous T-lymphocytes
(Table II). However, M.leprae granuloma macrophages failed to enhance this response. Neither mononuclear cell populations were able
to act as accessory cells for antigen (tuberculin) induced T-cell
proliferation. In these studies both epithelioid cells and macrophages were prepared in a purified state on a fluorescence activated
cell sorter using a specific monoclonal anti guinea pig macrophage
antibody. The failure of BCG induced epithelioid cells to support
tuberculin induced proliferation could be due to their lack of Ia
antigens. However, Kammer and Unanue [11] have shown that Ia -ve
macrophages can act as accessory cells for Con A induced proliferation, despite Ia positively being necessary for proliferative responses to antigen. The lack of response when M.leprae granuloma
macrophages, which are Ia positive, are used could be due either to
an inability to secrete interleukin-1 (IL-1) or to the secretion of

macrophages suppressor factors that might inhibit the secretion of IL-2. A range of accessory cell lymphocyte ratios were used in this study to exclude the possibility that a failure of cell response was due to suboptimal or supraoptimal accessory cell:lymphocyte ratios.

Table II. ACCESSORY CELL FUNCTION OF BCG EPITHELIOID CELLS AND M.LEPRAE MACROPHAGES IN CON A LYMPHOCYTE PROLIFERATION (cts/min)

L + 3 μg Con A	284 ± 200	62 ± 32
L + A	170 ± 26	44 ± 11
L + A + 3 μg Con A	16269 ± 4029	44 ± 5

PERITONEAL MACROPHAGES (CONTROL)

L + 3 μg Con A	175 ± 3	60 ± 51
L + A	594 ± 64	294 ± 8
L + A + 3 μg Con A	16318 ± 4542	15092 ±1605

L = Lymphocytes
A = Accessory Cells
Con A = Concanavalin A

REFERENCES

1. R. Virchow, Die Krankhaften Geschwulste. August Hirschwald Berlin, (1865).
2. J. Cohnheim, Vorlesungen uber Allgemeinen Pathologie, August Hirschwald, Berlin, (1877).
3. E. Metchinkoff, Lectures on the Comparative Pathology of Inflammation, Kegan Paul, Trench, Trubner, London, (1983).
4. H. Zinsser, Bacterial allergies and tissue reactions. Proc. Soc. exp. Biol.Med., 22:35, (1825).
5. J.L. Turk and D. Parker, Sensitization with Cr. Ni and Zr salts and allergic type granuloma formation in guinea pigs. J. invest. Derm. 68:341, (1977).
6. J.L. Turk, P. Badenoch-Jones and D. Parker, Ultrastructural observations on epithelioid cell granulomas induced by Zirconium in the guinea pig. J. Path. 124:45, (1978).
7. R.B. Narayanan, P. Badenoch-Jones and J.L. Turk, Experimental mycobacterial granulomas in guinea pig lymph nodes: ultra-structural observations. J. Path. 134:253, (1981).
8. R.C. Mathew, I. Katayama, S.K. Gupta, J. Curtis and J.L. Turk, Analysis of cells of the mononuclear phagocyte series in experimental mycobacterial granuloma by monoclonal antibodies. Infect. and Immun. 39:344, (1983).

9. R.B. Narayanan, P. Badenoch-Jones, J. Curtis and J.L. Turk.
   Comparison of mycobacterial granulomas in guinea pig lymph
   nodes, J. Path. 138:219, (1982).
10.R.B. Narayanan, J. Curtis and J.L.Turk, Release of soluble
   factors from lymph nodes containing mycobacterial granulomas
   and their effect on fibroblast function in vitro. Cell Immunol.
   65:93, (1981).
11.G.M. Kammer and E.R. Unanue, Accessory cell requirements in
   the proliferative response of T-lymphocytes to hemocyanin.
   Clin.Immunol. Immunopathol. 15:434. (1980).

THE ROLE OF COLLAGEN IN HEALING WOUNDS

A. Shuttleworth

Department of Biochemistry
Manchester University Medical School
Stopford Building, Oxford Road, Manchester M13 9PT

The restoration of tissue continuity following injury inevitably involves connective tissue proliferation. The time dependent sequence of events leading to the formation of scar tissue involves the interplay of many cell types and appears to follow a similar course of events irrespective of the site of injury in the body. Scar tissue is desirable from the standpoint of restoring the physical integrity of the tissue, but can be extremely disabling and may even produce lethal complications if it interferes with function in a vital organ.

Collagen is the major extracellular fibrous protein and its deposition in a healing wound can be correlated with gains in tensile strength. Most experimental investigations of repair have been carried out on wounds of skin because of its accessibility, and, because of its preponderance many of these investigations have been concerned with the nature and amount of collagen in repair tissue.

Collagen

Tissue collagens are seen in the electron microscope as long transversely banded fibrils which appear discrete. The fibril diameter is known to vary considerably in any one species from tissue to tissue, and in any one tissue to change with age. These changes in collagen fibril mass average diameters correlate positively with the tensile strength of the tissues [1]. While the term collagen originally referred to the cross-striated fibrous elements seen in appropriately stained sections, in recent years it has become apparent that it describes a family of closely related proteins [2].

211

To date 10 genetically distinct collagens have been described (see Table 1), and while the biological requirement for all these collagens is not understood, it serves to show that collagen has other than structural roles in the body.

The characteristic features of the chains making up the various collagen molecules is the presence of repeating Gly-X-Y triplets in which the X and Y positions are frequently occupied by prolyl and hydroxyprolyl residues respectively (Figure 1.). In addition, hydroxylysine residues are found in the Y position. These residues are important, not only for the glycosylation of collagen, but because they may participate in intermolecular cross-linking.

Variations in amino acid composition may occur in the various collagens but the Gly-X-Y triplet is essential for the triple helical conformation. In type I collagen the individual polypeptide chains each contain approximately 1000 amino acid residues while the tropocollagen molecule contains three polypeptide chains twisted around each other like a three-stranded rope. The resulting molecule is long (3000Å) and narrow (15Å) (see Figure 1.).

In the extracellular space the tropocollagen molecules spontaneously aggregate to produce fibrils with characteristic cross-striations repeating every 640Å. This repeat distance of the fibril structure is designated D (Figure 1.). Subsequently covalent intermolecular cross-links form between adjacent molecules stabilising the fibrillar assembly. For a detailed review of collagen structure the reader is referred to the many excellent reviews that are available [10,11,12].

In addition to the repeating triplet, collagen chains contain non-triplet sequences which instead of assuming triple-helical conformation assume a globular shape. These sequences may be at the ends of the triple helix, characteristically seen in procollagens I, II and III, or may intersperse short helical segments (type VI and IX),

The form and structure of collagen fibrils is also dependent on the nature of the collagen. This is clearly seen in type IV collagen which is not processed and contains globular domains which produce end-to-end contacts and give an open three-dimensional network, in stark contrast to the precise axial and lateral aggregation of type I collagen monomers.

Collagen fibril formation requires rod-like molecules, and the interstitial collagens I, II and III clearly fulfil this requirement. Differences in fibril diameter between these collagen types presumably reflect subtle changes in amino acid sequences, although specific interactions with other matrix components cannot be ruled out. Likewise, the changes that occur in type I collagen

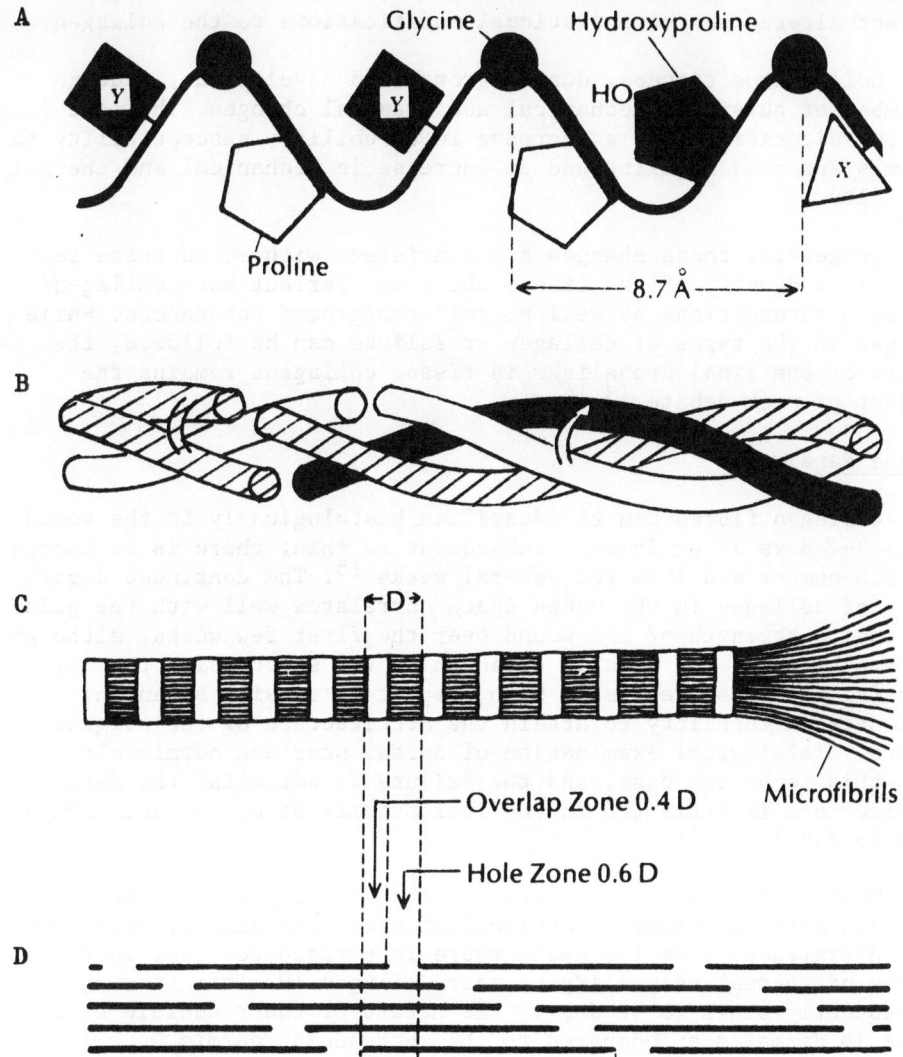

Fig. 1.   Schematic representation of the structure of the collagen
          fibril.
          A - amino acid sequence in collagen α-chain showing
          glycine occurs every third position and presence of large
          amounts of proline and hydroxyproline.
          X and Y represent any amino acid other than glycine.
          B - tropocollagen molecule contains three polypeptide
          chains associated in a triple-helical conformation.
          C - stained fibril of collagen exhibiting regular repeat
          period (D) of approximately 640Å.
          D - quarter-staggered packing of tropocollagen molecules.

fibril diameter between tissues, and in one tissue with age, may reflect altered post-translational modifications to the collagen.

Collagenous tissues, during growth and development, undergo a number of physical, mechanical and chemical changes. The most obvious alterations are a decrease in solubility, susceptibility to enzymes and acid swelling and an increase in mechanical and thermal stability.

In general these changes are consistent with an increase in cross-link density of the tissue which may reflect both collagen/collagen interactions as well as collagen/ground substances. While changes in the types of collagen crosslinks can be followed, the nature of the final crosslinks is tissue collagens remains the subject of much debate [13,14].

Tissue Repair

Collagen fibres can be identified histologically in the wound space 3-4 days after injury, subsequent to this, there is an increase in both number and size for several weeks [15]. The continued deposition of collagen in the wound space correlates well with the gain in tensile strength of the wound over the first few weeks, although it never aggains the tensile strength of the surrounding tissue. This failure of scare tissue with respect of tensile strength, indicates an inability to attain the architecture of the original tissue. Histological examination of dermal scar and dermis clearly show this to be the case, and the failure to establish the dermal architecture is reflected in the inelasticity of dermal scar compared to dermis [16,17].

While morphological studies clearly show differences between skin and scar, a number of chemical studies have also clearly shown this difference. From 1-3 weeks there is a rapid decrease in the degree of thermal solubility of scar collagen [18]. This finding was subsequently shown to be due to the nature of the reducible crosslinks in dermal scar compared to the surrounding dermis.

The predominant crosslink found in early scar collagen is hydroxylysino-5-keto-norleucine while that in dermis is dehydro-hydroxylysino-norleucine (Figure 2.) [19]. This difference in crosslinks accounting for the increased thermal stability of scar collagen compared to the surrounding dermis, since keto-imines are heat stable whereas Schiff's bases are thermally labile [20].

Changes in scar collagen crosslinking are seen with time and appear to mirror the changes seen in the development of dermis from embryonic to mature [21]. The significance of these changes is unclear. In the early acidotic conditions found in scar the increased acid stability of the keto-imine crosslink may be important in stabilising

COOH   OH   COOH

CH-CH$_2$ CH$_2$ CH-CH$_2$ N = CH-CH$_2$-CH$_2$-CH$_2$-CH

NH$_2$ NH$_2$

Dehydro-Hydroxylysino-Norleucine

COOH   OH   O   COOH
                                 "
CH-CH$_2$-CH$_2$-CH-CH$_2$-NH-CH$_2$-C-CH$_2$-CH$_2$-CH

NH$_2$ NH$_2$

Hydroxylysino-5-Keto-Norleucine

Fig. 2. Major reducible crosslinks present in dermal scar and
         skin collagen.

the collagen fibrils under these non-physiological conditions.
Under physiological conditions the rducible crosslinks, whether
Schiff's base or keto-imine, are stable and no change in tensile
properties of fibrils would be expected. Indeed, the change in
reducible crosslinks may not be of biological significance other than
that the extent of collagen lysine hydroxylation decreases generally
with maturation of tissues.

A number of investigations into the nature of repair collagen
have also indicated alteration in the collagen type [22,23,24]. Thus
the early stages of repair are characterised by elevated levels of
type III collagen which fall with time [22,23], type I collagen is,
however, the major collagenous species [25,26].

It has been suggested that the changes that occur during the
formation of a mature scar, follow those that occur during development of skin [22]. Thus, the changes in lysine hydroxylation and
cross-linking pattern in dermal scar parallel those that occur in
dermal development. Similarly, the fall in type III collagen seen
in dermal scar is reflected in the changes in type III in dermis
pre- and post-natally [27,28]. However, this comparison is qualitative
rather than quantitative, since the amount of type III collagen in
dermal scar is always less than that in foetal skin.

The role of type III collagen in tissues is unknown, although
it is always associated with type I collagen. In contrast to type I
the conversion of type III procollagen is slow and this may limit
the fibril diameter found in tissues. Fibrils from type III collagen
molecules are present in the more distensible connective tissues,

blood vessels, periodontal ligament, uterine wall and while in some cases they are associated with elastic fibres, in others they surround type I collagen fibres. Whether they provide the means whereby the larger type I fibres and fibrils are able to slide past one another and accommodate elastic pressures, remains to be elucidated.

The elevation of type III collagen in developing tissues and in the early stages of repair suggests a role in the formation as well as contributing to the structural integrity of the tissue. Its persistence, however, in several adult tissues suggests a more specific role in tissue organisation and function. Its presence in several elastin-rich tissues and its close association with elastic fibres, not type I collagen fibres, supports this view.

Increased levels of type III collagen relative to the surrounding tissue is also seen in many foreign body reactions. Replacement of cruciate ligament in sheep with carbon fibres leads to ingrowth and development of new tissue in which the amount of type III collagen is higher than in control ligaments and which decreases with time [29]. In a study on human bone fractures where there had been abnormal healing, an increased amount of type III collagen was found, in some cases accounting for 40% of the total collagen in the tissue [30]. This increase in type III collagen presumably being a reflection on the nature of the fibrous tissue lying between the bone ends.

Implanted materials would similarly be expected to elicit a similar response, although the amount of type III found in the developing fibrous capsule would presumably vary depending upon the time of sampling.

That inflammation is an obligatory step in repair is well documented, how this inflammation stimulates fibroblast replication is, however, not known. Many inflammatory lesions are associated with collagen destruction [31, 32], and one should be cautious about interpreting results based on the proportion of a particular collagen type, since it has been shown that in chronic periodontitis, that while the proportion of type III collagen changed, the amount remained the same [33]. This would imply that in the chronic inflammatory situation the cellular response has been altered. Interestingly in this situation, it was the amount of type V collagen which was increased, although the pathological and functional significance of this is unclear.

The major events of the sequence of repair are well known, injury leads to inflammation, fibroplasia and finally a scar. These series of events are characterised by the appearance of specific cell types, ultimately leading to the appearance of a new extracellular matrix. While in some cases wounds may fail to heal properly because of altered collagen synthesis, the more common problem is overproduction of collagen and formation of fibrous scars. Excessive fibrosis is a major problem in man, and its control will be a major achievement. The knowledge acquired from our understanding of collagen biochemistry will hopefully enable realisation of this goal.

TABLE 1

Distribution and Molecular Configuration of Collagen Types

Collagen Type	Chain Composition	Distribution
I	$[\alpha_1(I)]_2 \alpha_2(I)$	Most tissues except cartilage [2]
I trimmer	$[\alpha_1(I)]_3$	Skin, tumour, dentine [2]
II	$[\alpha_1(II)]_3$	Cartilage, vitreous [2]
III	$[\alpha_1(III)]_3$	Developing and Distensable tissues [2]
V	$[\alpha_1(V)]_2 \alpha_2(V)$	Pericellular-fibroblasts and smooth muscle cells [2]
IV	$[\alpha_1(IV)]_2 \alpha_2(IV)$	Basement membranes [3]
VI	$(140K)_3$?	Placenta, skin aorta [4]
VII	$[\alpha_1(VII)]_3$	Anchoring fibrils [5]
VIII	175K	Normal and malignant cells in vitro [6]
IX	$\alpha_1(IX) \alpha_2(IX) \alpha_3(IX)$	Cartilage [7]
X	$[\alpha_1(X)]_3$	Cartilage-growth plate [8]
XI ?	$[1\ 2\ 3]$	Cartilage [9]

REFERENCES

1. D.A.D. Parry, A.S. Craig and G.R.G. Barnes, In: "Fibrous Proteins:
    Scientific, Industrial and Medical Aspects". D.A.D. Parry
    and L.K. Creamer, ed: Academic Press, London Vol. 2,
    77-88 (1979).
2. S. Gay and E.J. Miller, Ultrastructural Pathology 4, 365-77
    (1983).
3. M.E. Grant, J.G. Heathcote and R.W. Orkin, Biosci. Rep. 819-42
    (1981)
4. S. Ayad, D.A. Shuttleworth and M.E. Grant Biochem. Soc. Trans.
    12, 1052-53 (1984).
5. H. Bentz, M.P. Morris, L.W. Murray, L.Y. Sakai, D.W. Hollister
    and R.E. Burgeson, Proc. Natl. Acad. Sci. USA 80, 3168-72,
    (1983).
6. H. Sage, B. Trueb and P. Bornstein, J. Biol. Chem. 258, 13391-
    13401 (1983).
7. S. Ayad, M.Z. Abedubm J. B. Weiss and S.M. Grundy FEBS Lett 139,
    300-304 (1982).
8. G.J. Gibson, C.M. Kielty, C. Garner, S.L. Schor and M.E. Grant,
    Biochem. J. 211, 417-26 (1983).
9. R.B. Burgeson and D.W. Hollister, Biochem. Biophys. Res. Commun.
    87, 1124-31 (1979).
10. P. Bornstein and W. Traub, In: Neurath. H. & Hill, R.L. (eds)
    "The Proteins" Vol IV Academic Press New York 411-32 (1979).
11. D.R. Eyre Science 207, 1315-22 (1980).
12. E.J. Miller, In: "Extracellular Matrix Biochemistry" ed. Piez,
    K.A. and Reddi A.H. Elsevier, N.Y. 41-82, (1984).
13. S.P. Robins, In: "Collagen in Health and Disease" ed. Weiss J.B.
    and Jayson M.I.V. 160-78, (1982).
14. D.R. Eyre, M.A. Paz and P.M. Gallop, Ann. Rev. Biochem. 53, 717-
    748 (1984).
15. R. Ross, Biol. Revs. 43, 51-84, (1968).
16. J.C. Forrester, B.H. Zederfeldt and T.K. Hunt, J. Surg.Res. 9,
    207-211 (1969).
17. J.C. Forrester, T.K. Hunt, T.L.O. Hayes and R.F.W. Pease,
    Nature 221, 373-74 (1969).
18. L. Forrest, J. Dixon and D.S. Jackson, Connect Tiss. Res. 1,
    243-50, (1972).
19. L. Forrest, C.A. Shuttleworth, D.S. Jackson and G. Mechanic,
    Biochem. Biophys. Res. Commun. 46, 1976-81, (1972).
20. N.D. Light and A.J. Bailey, In: Fibrous proteins: Scientific
    Industrial and Mechanical Aspects. Parry D.A.D. and Creamer
    L.K. ed. Academic Press, London Vol.1, 151-77, (1979).
21. A.J. Bailey and S.P. Robins, In: "Frontiers of Matrix Biology",
    Robert L. ed. 130-35, (1973).
22. A.J. Bailey, S. Bazin, T.J. Sims, M. Lelous, C. Nicoletis and
    A. Delauny, Biochem. Biophys. Acta 404, 412-421 (1975).
23. M.J. Barnes, L.F. Morton, R.C. Bennett, A.J. Bailey and T.J. Sims,
    Biochem. J. 157, 263-68,(1976).

24. J.N. Clore, I.K. Cohen and R.F. Diegelmann, Proc. Soc. Exp. Biol. & Med. 161, 337–39, (1979).
25. C.A. Shuttleworth and L. Forrest, Biochem. Biophys. Acta. 365, 454–57, (1974).
26. A.J. Bailey, T.J. Sims, M. Lelous and S. Bazin, Biochem. Biophys. Res. Commun. 66, (1975).
27. C.A. Shuttleworth and L. Forrest, Europ. J. Biochem. 55, 391–95, (1975).
28. E.H. Epstein, J. Biol. Chem. 249, 3225–3231, (1974).
29. I.M. Forster and C.A. Shuttleworth, J. Bone Jt. Surg. Submitted (1984).
30. A.M. Anderson, G.W. Hastings, T.R. Fisher, E.R.S. Ross and A. Shuttleworth, J. Bone Jt. Surg. Submitted (1984).
31. E.D. Harris, Arthr. Rheum. 19, 68–72, (1976).
32. R.C. Page and H.E. Schroeder, Inter. Dent. J. 23, 455–69, (1973).
33. A.S. Narayanan, L.D. Engel and R.C. Page, Coll.Rel.Res. 3, 323–34, (1983).

THE RABBIT EAR CHAMBER: A STUDY OF CELL INTERACTIONS

D.J. Leaper, M.E. Foster, S.S. Brennan and I.A. Silver

University Departments of Surgery and Pathology
University of Bristol, Bristol, UK

INTRODUCTION

The rabbit ear chamber was designed by Sandison (1) to allow dynamic microscopic observations of cells and tissues in living animals. His original chambers were made of two rectangular pieces of isinglass connected together to form a central observation area. Many biologists have since used the ear chamber to study infection, in particular TB granulomas, (2) tumour growth in relation to tumour kinetics (3) and the tissue reaction to implanted prosthetic materials (4). Silver (5) has examined the healing process in detail using his own modifications of the rabbit ear chamber. He found that the process of repair by the module of the macrophage, fibroblast and angioneogenesis (healing by secondary intention) was highly dependent on gradients of tissue oxygenation, pH and glucose. Collagen formation by fibroblasts, the end result of successful repair, was found to be limited by a narrow range of tissue PO_2 in these studies.

We have investigated the interaction of commonly used surgical materials and antiseptics with cells in the rabbit ear chamber.

THE RABBIT EAR CHAMBER

The rabbit ear chamber used in these studies is shown in figure 1. It consists of polycarbonate body and central plug which contains a transparent cover-slip which allows direct microscopic observation of a healing wound. The granulation tissue of the healing process can be monitored as it slowly fills the dead space of the chamber through the eight circumferential

221

holes. The dead space is usually filled by mature granulation, in the absence of infection, by 4-6 weeks (figure 2).

Under sterile conditions, and antibiotic prophylaxis, ear chambers were inserted into the ears of anaesthetised lop-eared rabbits. The ears were prepared by hair removal and skin disinfection with 1% Hibitane. The skin on the inner surface of the ear was raised by gentle blunt dissection to produce a small pocket. By punching a 6,25 mm hole through the ear cartilage, positioned at the centre of the skin pocket, the chamber could be positioned and the Marlex skirt enclosed within the pocket. Chambers were left for 4-6 weeks to mature.

IMPLANTATION OF TEST MATERIALS

The coverslip of the chamber was cleaned and flooded with phosphate-buffered-saline (PBS) and penicillin solution. The Circlip was removed together with the Teflon spacer and the cover-slip lifted (see figure 1). The test material was then placed onto the tissue surface and a new coverslip, with a 2 mm eccentric hole, placed over the chamber and a second ordinary coverslip replaced. Photographs were taken serially of the tissue reaction using a dissecting microscope and an Olympus OM camera.

i) Suture materials. 2 mm lengths of catgut, silk and nylon were inserted into separate chambers and observed over a 4 week period. Figures 3-5 show the different tissue reactions at 3 days. Catgut invoked an early profound tissue reaction with oedema, capillary dilatation and an exaggerated acute inflammatory response. The reaction to silk was similar but less marked with separation of the braided silk fibres whereas nylon appeared to be totally inert.

ii) Dressing materials. Four dressing materials currently popular for dressing open wounds, ulcers and burns were evaluated: pig-skin collagen, polyacrylamide gel, polyurethane foam and silastic elastomer foam. Each material was inserted into a separate ear chamber and observed over four weeks. The tissue reactions seen at approximately one week are shown in figures 6-9. In the ear chamber containing the pig-skin collagen dressing new vessels were seen growing through the collagen matrix. Polyurethane and silastic foam invoked a mild tissue reaction but polyacrylamide gel was the least irritant.

REACTION TO ANTISEPTICS

Mature chambers were flooded with 0.5 ml of PBS (control), Eusol or aqueous 0.05% Chlorhexidine and photographs were taken over the following seven days. After the hypochlorite solution, Eusol, there was a complete cessation of blood flow, which did

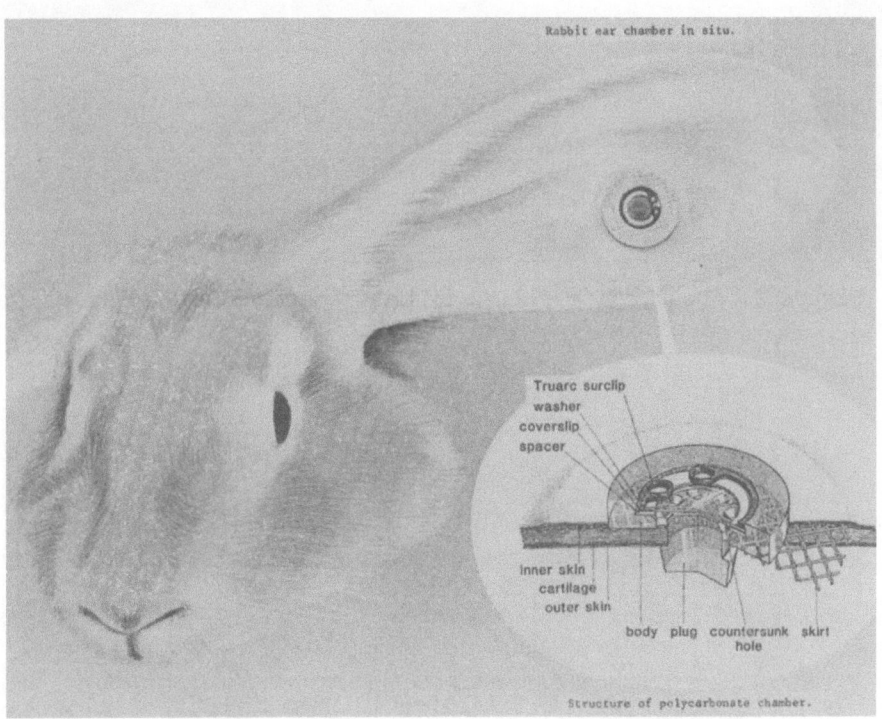

Figure 1.   The rabbit ear chamber in situ.

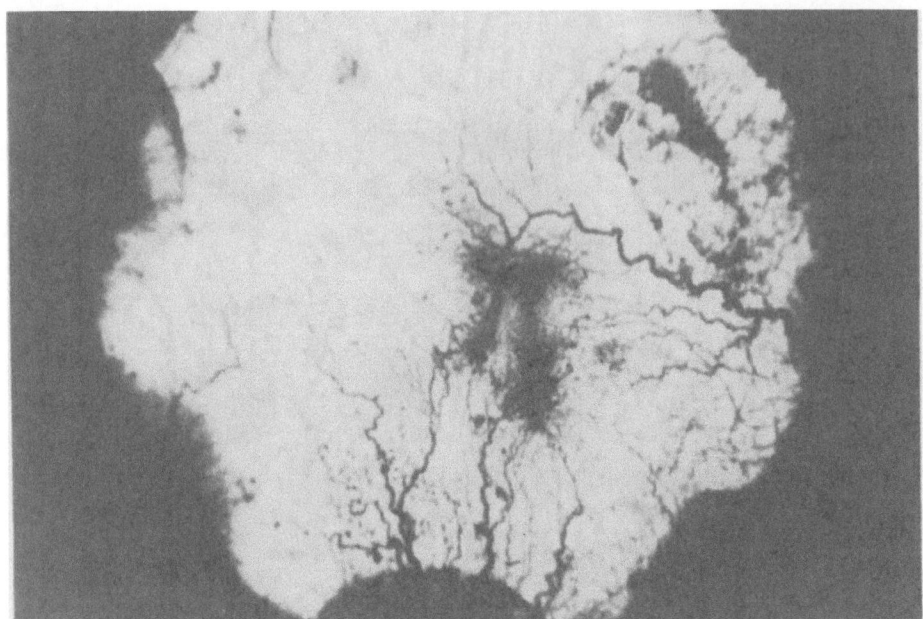

Figure 2.   Dead space filled by mature granulation tissue

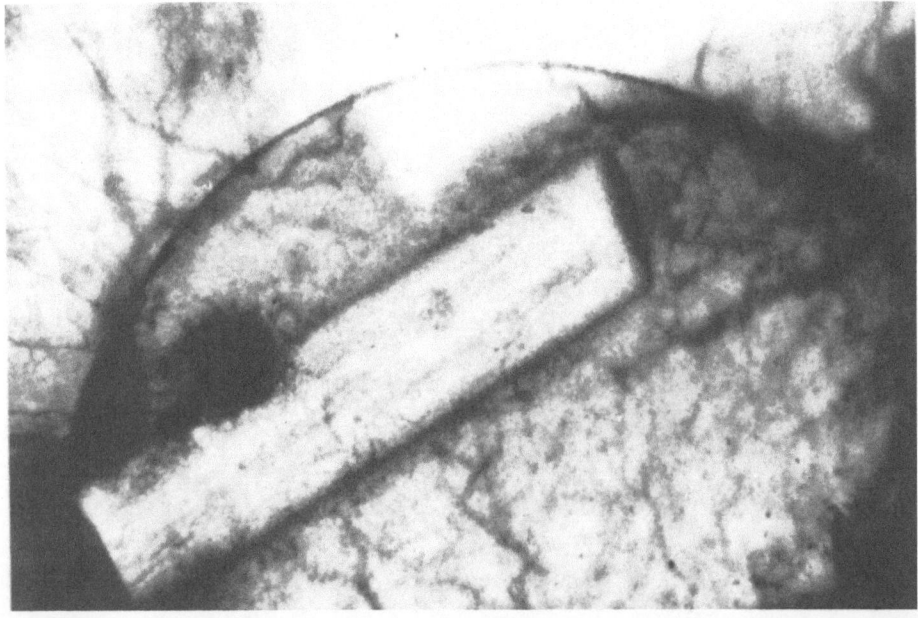

Figure 3.   3-day reaction to catgut. Oedema, dilatation and extra
            new vessels in exaggerated acute inflammatory response

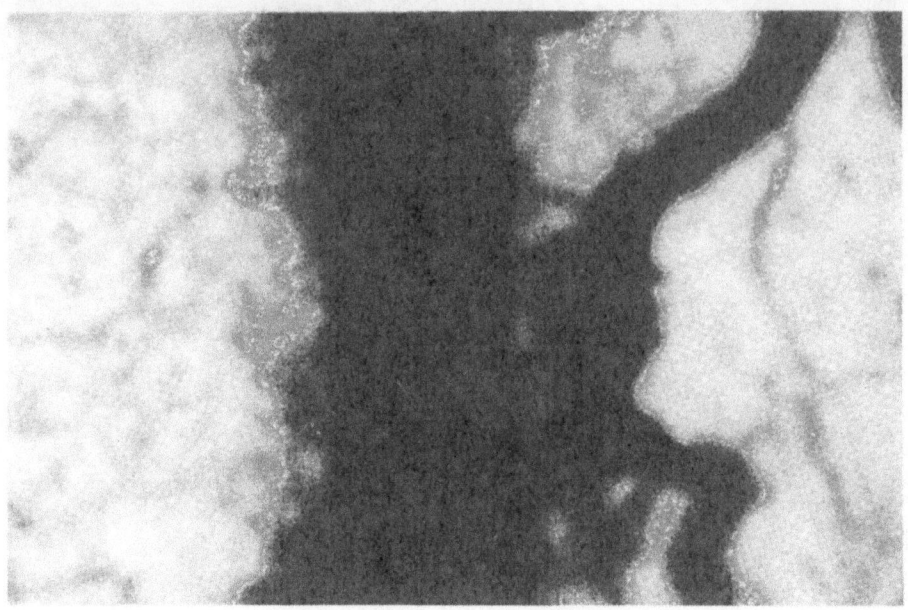

Figure 4.   3-day reaction to silk. Similar reaction as catgut with
            separation of braid

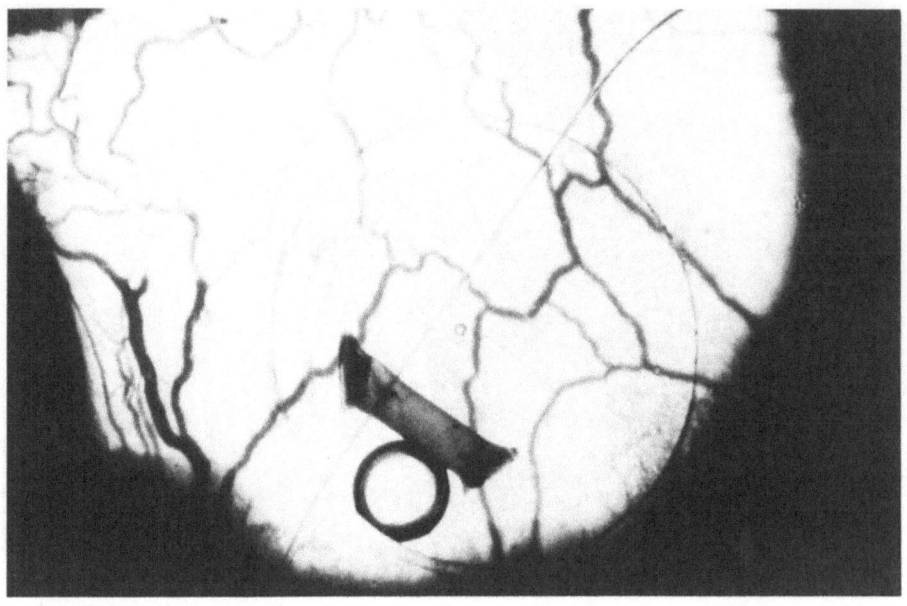

Figure 5.   3-day reaction to nylon. Suture totally inert

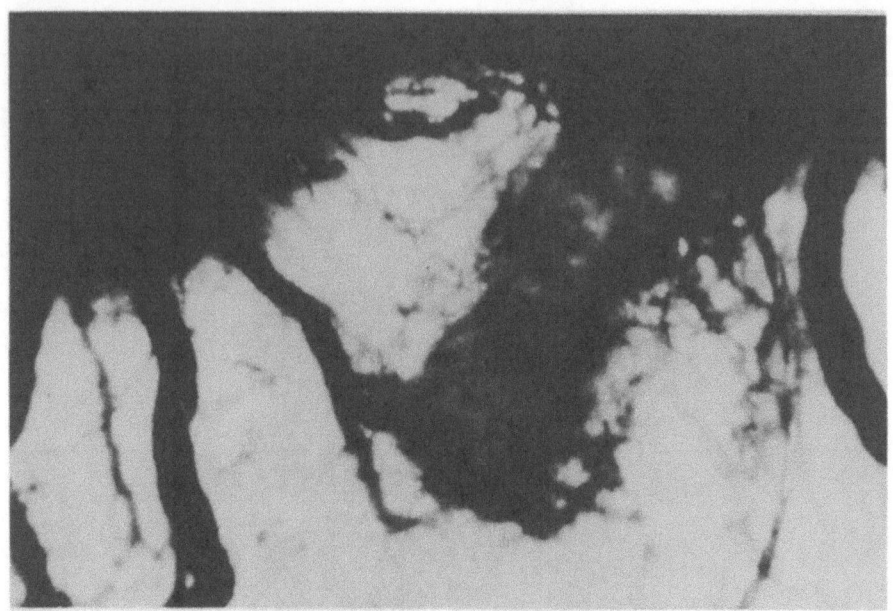

Figure 6.   1-week reaction to pig-skin collagen. New vessels seen
            invading the collagen dressing

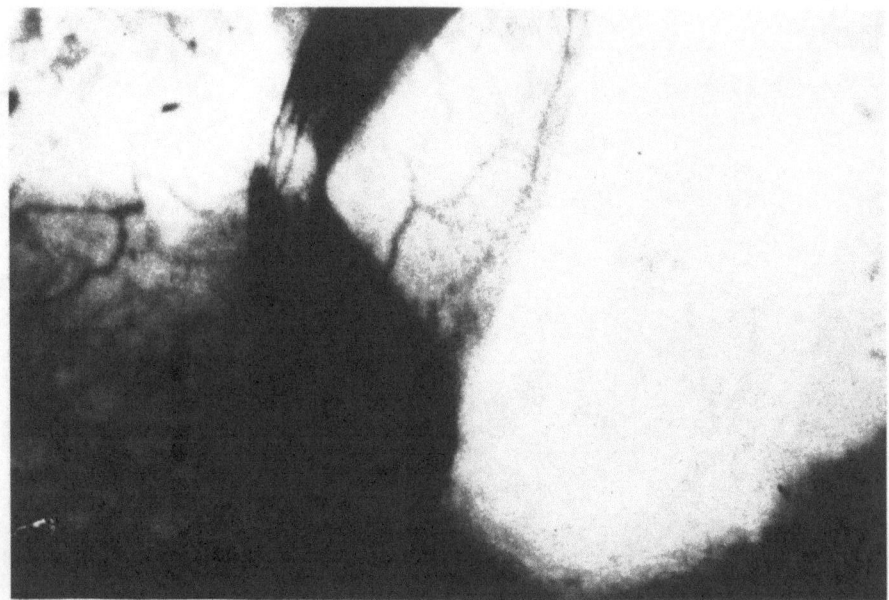

Figure 7.   1-week reaction to Polyacrylamide gel (Geliperm).
            Edge of dressing in centre of picture shows inert
            reaction at the edge

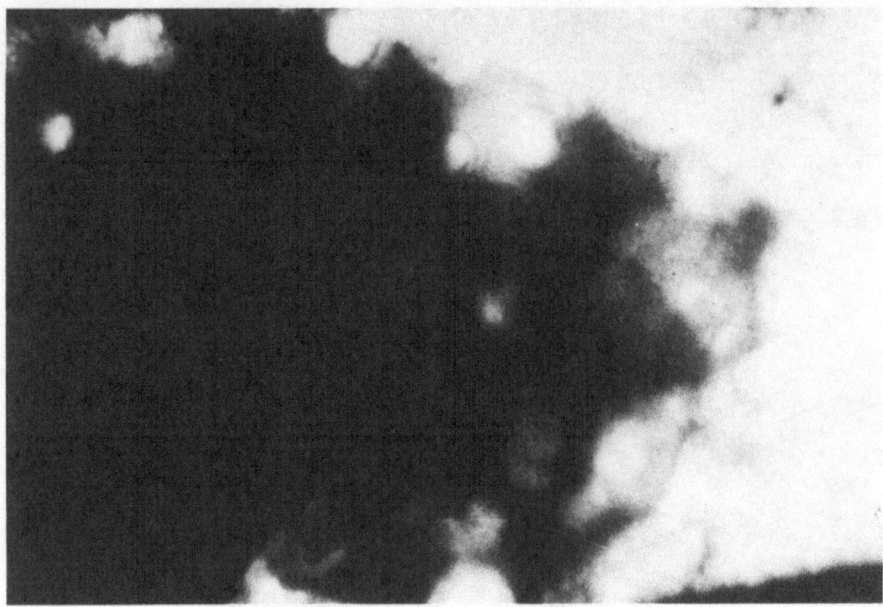

Figure 8.   1-week reaction to Polyurethane foam (Lyofoam). Mild
            cellular reaction with vasodilatation around foam

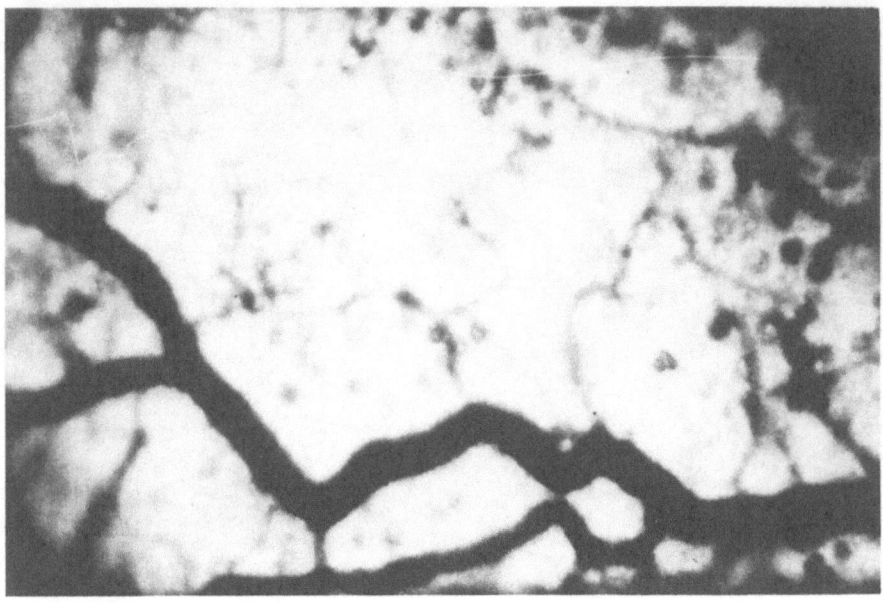

Figure 9.   1-week reaction to silastic foam elastomer.
            Capillary dilatation only

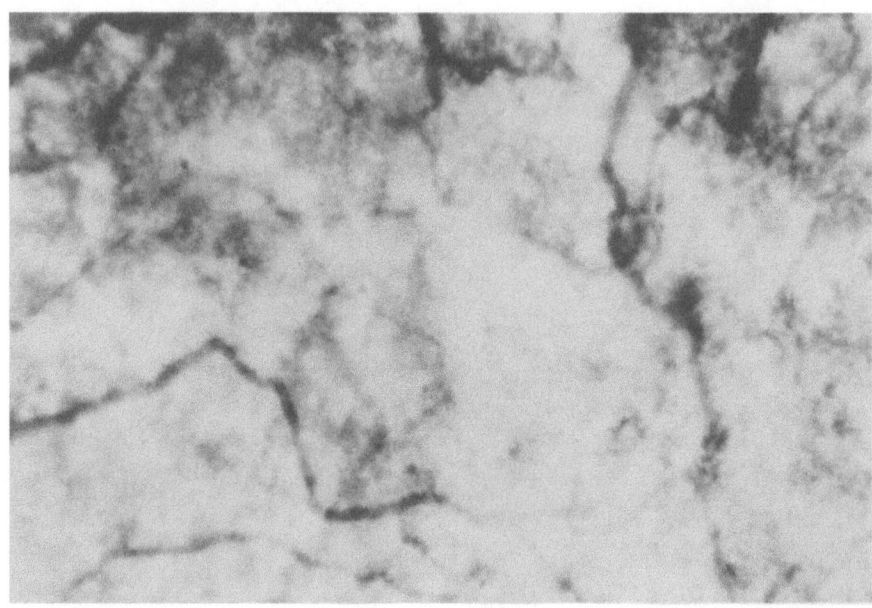

Figure 10. Immediate response to Eusol. Complete cessation of
            blood flow with capillary shutdown and acute
            inflammatory response with exudate

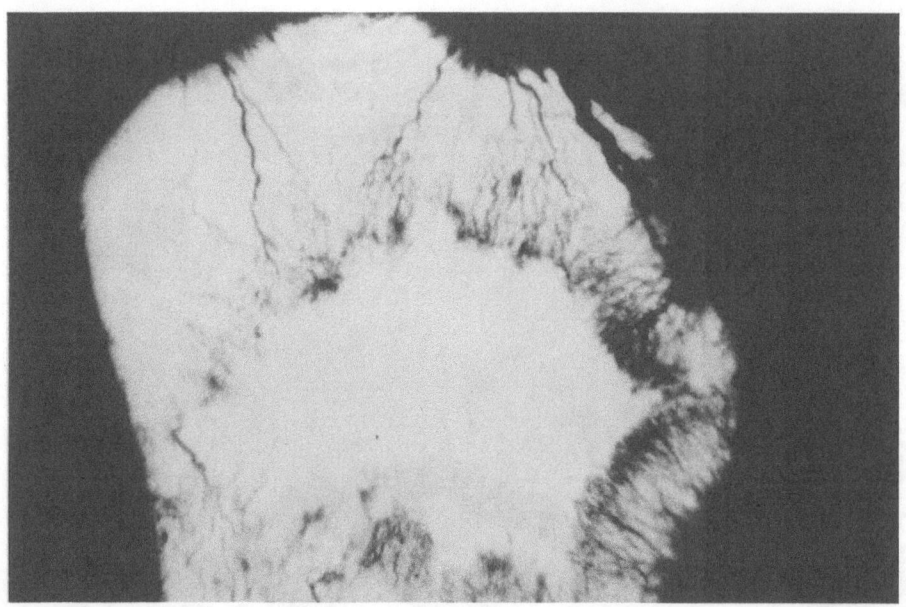

Figure 11. 5-day response after Eusol. Ghosts of old capillaries
            centrally with ingrowth of new granulation tissue
            peripherally

not recover, with an attendant acute inflammatory response
(figure 10). This was followed by death of all elements of the
granulation tissue with the regrowth of new vessels and granulation
tissue 5 days later (figure 11). Chlorhexidine caused a mild
exudative reaction with a temporary redirection of blood flow into
larger blood vessels but PBS caused no demonstrable changes.

The rabbit ear chamber has proved to be a useful biological
instrument for dynamic in vivo evaluation of the biocompatibility
of surgical materials and dressings. The exaggerated tissue
response to catgut confirms the current general opinion amongst
many surgeons that it should be replaced by the less irritant,
more predictable, synthetic monofilament sutures. There is an
ever-increasing number of surgical dressing materials being
introduced onto the market and these studies using the rabbit ear
for evaluation of their compatibility in vivo may be helpful in
determining whether an inert or a biologically active dressing is
required. The current use of some antiseptics may be harmful to
the healing process and the use of hypochlorite solutions in
particular, although active against micro-organisms colonising a
healing ulcer, may well stop or damage formation of healthy granulation tissue.

REFERENCES

1. J.C. Sandison, A new method for the microscopic study of
        living growing tissues by the introduction of a trans-
        parent chamber in the rabbit's ear. Anat. Rec. 28 :
        281-87, (1924).
2. R.H. Ebert and W.R. Barclay, The effect of chemotherapy on
        the tissue response to tuberculous infection as observed
        in vivo. J. Clin. Invest. 29 : 810, (1950).
3. S. Wood, Pathogenesis of metastasis formation observed in vivo
        in the rabbit ear chamber. Arch. Pathol. 66 : 550-67.
        (1958).
4. G.F. Howden and I.A. Silver, The use of an improved rabbit
        ear chamber technique for the study of dental materials.
        International Endodontic J. 13 : 3-16, (1980).
5. I.A. Silver, The physiology of wound healing. In: Wound Healing
        and wound infections, Hunt T.K. ed., Appleton-Century-
        Crofts. New York. (1980).

THE TREATMENT OF SKIN GRAFT DONOR SITES

Keith Poskitt, Edward Lloyd-Davies, Alan James and
Charles McCollum

Department of Surgery
Charing Cross and Westminster Medical School
Fulham Palace Road, London W6 8RF

Skin grafting is commonly used in the treatment of burns, venous
and ischaemic ulceration and post-operative wound defects. Most skin
grafting can adequately be performed with the use of free skin grafts
which can be broadly categorised into whole and split skin grafts.
The use of pinch skin grafts (Fig 1) which are in effect whole skin
grafts centrally and split skin grafts peripherally are more
frequently being used for the treatment of venous ulceration[1,2].
It is well described that the donor sites from which skin grafts are
obtained are the source of more discomfort than the recipient site.
This discomfort is caused by exposure of the dermal nerve endings
and is diminished more rapidly if epithelialisation occurs readily.
The presence of infection delays healing and can in addition cause
increased discomfort. Donor site discomfort may also vary depending
on the site from which the skin graft is obtained. In pinch skin
grafting the donor sites more commonly used are the anterolateral
and anteromedial aspect of the thigh.

When cosmesis is important in younger patients pinch grafts 4
to 5 mm in diameter may be obtained in a linear fashion with subsequent excision and repair by subcuticular suture. In elderly
patients however, cosmetic appearance in this region may be less
important and the area is allowed to heal without excision and
results in some scarring.

Methods of treating skin graft donor sites aim at obtaining a
pain free, uninfected donor site with minimal changes of dressing.

231

More recently various occlusive dressings have become available
for the treatment of skin defects[3,4]. Re-epithelialisation is thought
to be increased by maintaining a moist environment in the presence
of proteins, hormones, enzymes and a variety of viable host cell types.
There is some evidence to suggest that tissue viability is best
maintained in the absence of dehydration. Migration of all cellular
elements in general and epithelial cells particularly is more efficient
and faster while oxygenation is optimised[4,5]. Occlusive dressings
which are impervious to water, micro-organisms and contamination
from without, while remaining selectively permeable to oxygen and
water vapour, have become available for use in skin defects. As an
adhesive semi-permeable polyurethane dressing (Fig 2) is theoretically
ideal for the treatment of skin graft donor sites, we performed a
prospective randomised study comparing the use of polyurethane
dressings to the previously used method of paraffin gauze dressings
for the treatment of pinch skin graft donor sites.

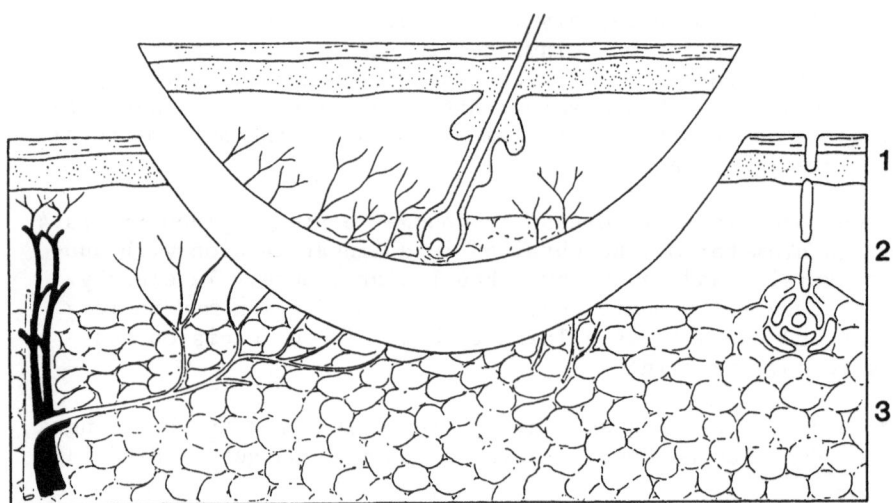

1   Epidermis

2   Dermis

3   Subcutaneous Fat

Fig. 1.   Diagrammatic representation of a pinch graft.

METHOD

In both treatment groups pinch skin grafts were obtained from the anteromedial aspect of the thigh following local infiltration with ½% Lignocaine. Following preparation of the skin with Chlorhexidine in alcohol skin was elevated with forceps and pinch skin grafts approximately 5 mm in diameter were excised with a No.

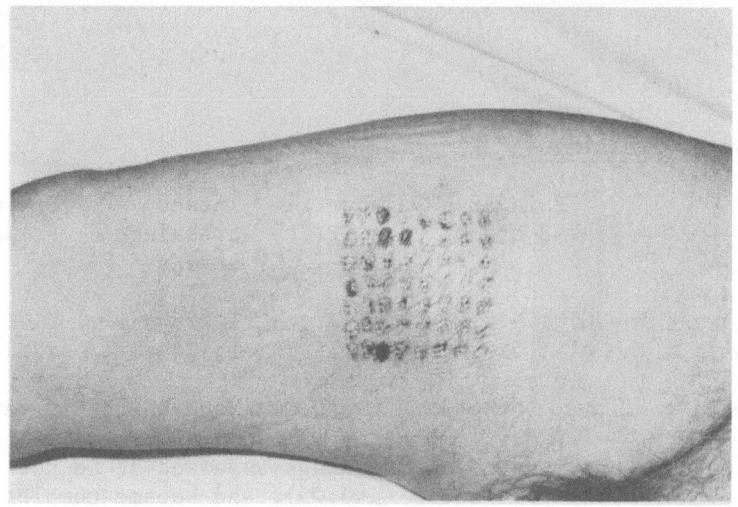

Fig. 2.   A polyurethane dressing placed over a pinch skin graft donor site.

20 scalpel blade. Pinch grafts were obtained in a uniform manner leaving 2 mm to 3 mm between each defect. The bleeding from the donor sites was controlled with direct pressure using a dry gauze for approximately 2 minutes. Following haemostasis the surrounding skin was dried and a dressing applied dependant on randomisation by selection from sealed envelopes. All donor sites were subjected to a standardised compression bandage including wool, crepe and elastic adhesive strapping.

Dressings were left in situ for 5 days with pain being assessed at 48 hours and the number of days until the donor site was pain free recorded. Pain was recorded on a numerical scale as below:-

1)   No discomfort
2)   Slight discomfort (aware of wound)
3)   Moderate discomfort (tolerable)
4)   Painful (requiring analgesia)

Inflammation and discharge were assessed at five days. The number of dressings that remained intact were recorded at day five. The comparison between the two forms of dressings was performed using the Mann-Whitney U test for non-parametric data.

RESULTS

Comparison of Dressings

	Pain at 48 hrs	Days until pain free	Intact dressings at day 5	Dry wounds at day 5
Paraffin gauze (n=26)	1.9±0.1	4.5±4.7	14/26	21/26
Polyurethane (n=24)	1.5±0.1*	1.8±0.5**	15/24	23/24

Results mean ± s.e.m.   *p<0.05   **p<0.01   Mann-Whitney U test

Patients with polyurethane dressings assessed at 48 hours had a significantly lower pain score (p<0.05), and became completely pain free earlier, than those with paraffin gauze dressings (p<0.01). The assessment of wound discharge at five days showed no significant difference in either study group. In both groups similar number of dressings remained intact at day 5.

In three cases in the polyurethane group excessive sero-sanguinous discharge accumulated beneath the polyurethane dressings (Fig 3) with subsequent leakage. In all cases this was associated with a large skin graft donor site involving over 30 individual pinch graft sites.

CONCLUSION

The polyurethane dressing performed at least as well as conventional paraffin gauze dressing with regard to patient comfort. There was no significant difference in the incidence of wound infection in either group or in the rate of wound healing. In all cases donor sites were completely healed within three weeks of surgery.

Polyurethane dressing does however have several distinct advantages over the conventional dressing. It is less costly, easier to apply and does permit direct inspection of wounds with minimal disturbance to the patient. Excessive sero-sanguinous discharge collecting beneath the polyurethane dressing is a definite problem but was confined only to the large donor sites in this study group. We would therefore recommend the use of polyurethane dressing for all but the larger pinch graft donor sites.

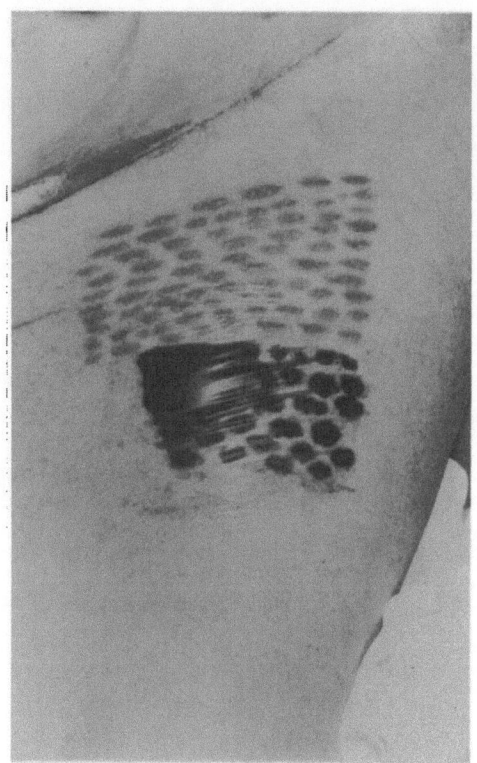

Fig. 3.   Collection of sero-sanguinous fluid beneath a polyurethane
           dressing in a large donor site at day 3.

REFERENCES

1.   A.S. Chilvers and G.K. Freeman, Outpatient skin grafting of
      venous ulcers, Lancet. 2:1087-1088 (1969).

2. L.G. Millard, M.M. Roberts and M. Gatecliffe, Chronic leg
    ulcers treated by the pinch graft method, Br. J. Dermatol.
    97: 289-95, (1977).
3. W.A. Gilmore and R.G. Wheeland, Treatment of ulcers on legs by
    pinch grafts and a supportive dressing of polyurethane.
    J. Dermatol. Surg. Oncol. 8: 177-83, (1982).
4. D.J. McCarthy and B. Montgomery, Polyurethane in the management
    of ulcerating lesions of the lower extremities, J.A.P.A.
    73: 1-9, (1983).
5. B.H. Fisher, Topical hyperbaric oxygen treatment of pressure
    sores and skin ulcers, Lancet, 2: 405-9, (1969).

THE EFFECT OF SURGICAL DRESSINGS ON SKIN TEMPERATURE

Carol Miller, Keith Poskitt and Ian Lane

Department of Surgery
Charing Cross Hospital
London W6 8RF

Infection of surgical wounds leads to delaying healing, incisional herniation and occasionally to fatal septic complications[1]. Contamination by micro-organisms at the time of surgery is prevented by an aseptic operating environment. Postoperatively, wounds are dressed to reduce entry of bacteria such as Staphyloccoci aureus and Eschericia coli which occur naturally as skin commensals.

Absorbant dressings act as thermal insulators and increase the surface temperature of skin to approach that of body core, $37^{o}C$. Most human pathogens are thermophilic and show optimum growth rate at $37^{o}C$[2]. Those dressings that increase skin temperature may lead to a higher incidence of postoperative infection. We have assessed the skin temperature adjacent to abdominal incisions under two commonly applied postoperative dressings.

PATIENTS AND METHODS

Forty one patients who had undergone laparotomy 2 - 12 (mean 4; sem 0.4) days previously were selected for entry into the study. Patients with infected wounds were excluded. Skin temperature was measured after allowing 10 minutes for equilibration by a thermocouple connected to a potentiometer (Light Laboratories, 3GID Instrument). Thermocouples were placed adjacent to the incision and at an identical position on the contralateral side of the abdomen to act as a control (Fig 1). Skin temperature was measured under no dressing and following the consecutive implacement of Airstrip (Smith & Nephew Limited) and Opsite (Smith & Nephew Limited) over the wound in all patients. Ambient air temperature, humidity and patient core temperature remained constant during each set of

237

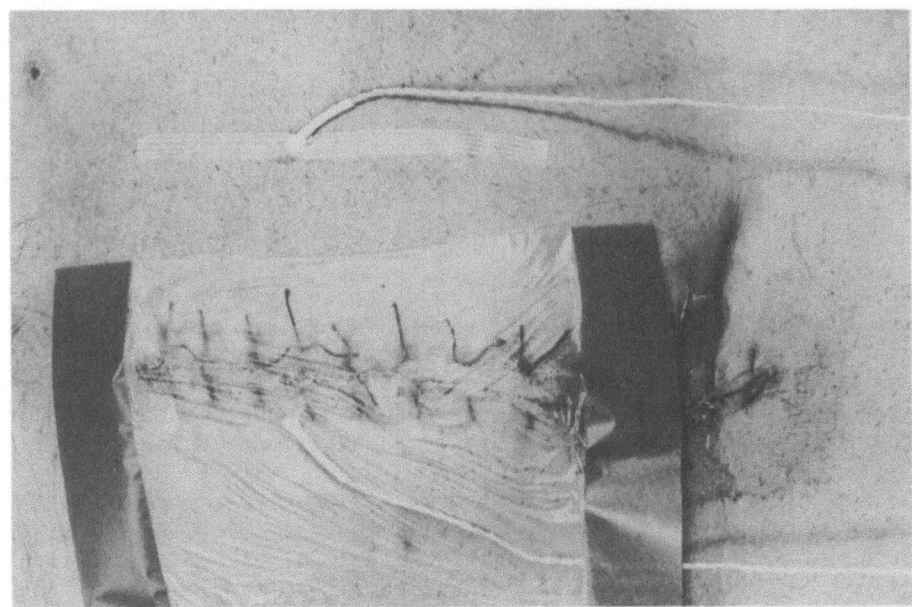

Fig 1   A laparotomy wound dressed with an elastomeric adhesive
         film showing a cutaneous thermocouple adjacent to the
         incision with the control on the contralateral side of the
         abdomen.

readings.   Results were analysed using the Wilcoxon signed rank test.

RESULTS

    Mean wound skin temperature recorded with no dressing was
34.9 ± 0.1 (sem) $^{\circ}$C. This increased to $35.1 \pm 0.1^{\circ}$C under Opsite
and $35.6 \pm 0.1^{\circ}$C under Airstrip (Fig 2). The skin temperature
achieved by Opsite was significantly lower than that obtained when
Airstrip was used as a dressing ($p < 0.01$).

    In order to assess the effects of these differing temperatures
on bacterial growth, broth containing Escherichia coli was incubated
simultaneously at 35°C and 36.5°C. The numbers of bacteria present
were assessed at 2, 4, 6, 8 and 24 hours following initiation of
the culture. At 24 hours the number of bacteria present per cubic
centimetre of broth at 36.5°C was over treble of the number found
at 35°C (Fig 3).

DISCUSSION

    Opsite is a semipermeable elastomeric adhesive film compared

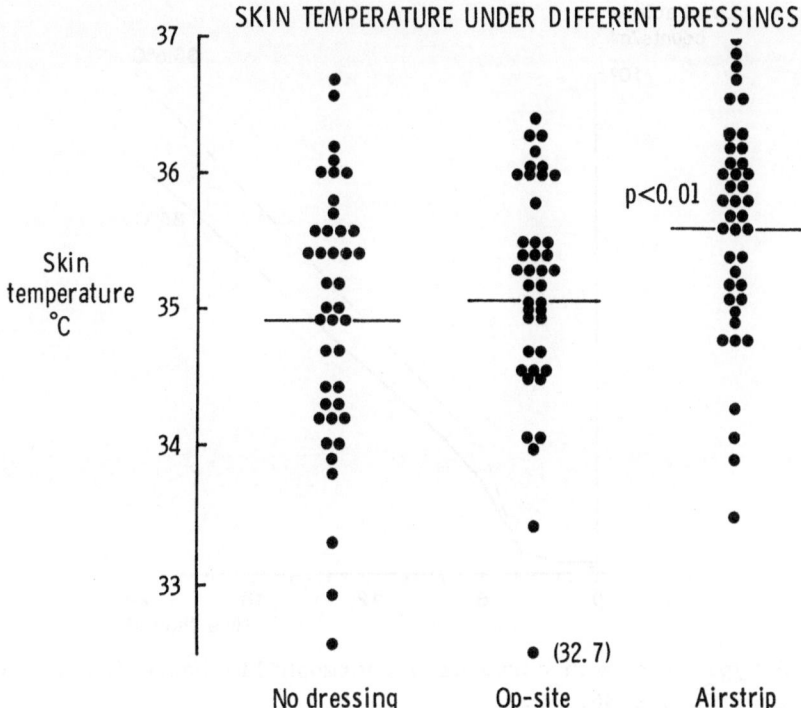

Fig. 2   Scattergram illustrating skin surface temperature achieved
with no dressing and under Opsite or Airstrip. (Results
show mean ± standard error of mean).

to the composite structure of Airstrip with a non-adhesive polymeric film on the inner side and a waterproof microporous plastic
external surface sandwiching a viscose acrylic mix. The significantly lower temperatures found with Opsite were shown to reduce the rate
of multiplication of Escherichia coli. It has been demonstrated
that following laparotomy, wound infections commonly originate from
skin surface organisms[3]. In addition to allowing constant observation of the wound, Opsite has the further advantage over absorbant
dressings of preventing excessive drying of the wound. It has been
shown in man and pigs[4,5] that epithelialisation is twice as rapid
in wounds kept moist compared to those desiccated by exposure to the
air. In addition, the oxygen permeable nature of the film will
allow epidermal cell migration to progress rapidly[6] and prevent the
phagocytic response to bacteria being overwhelmed with the subsequent
development of overt sepsis[7]. However, copious serous exudate will
collect under adhesive dressings and may become secondarily infected[8].
Epithelial growth may be delayed if it is necessary to change an
adherent dressing which may damage the developing neoepithelium.

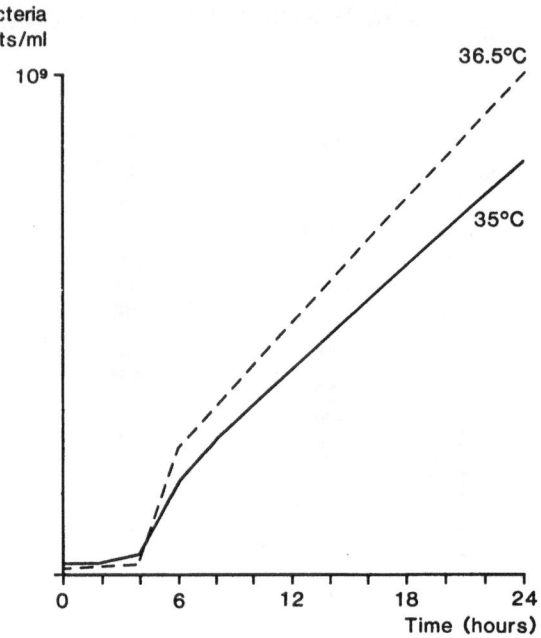

Fig. 3   A typical growth curve of a thermophilic bacteria, E. coli,
         at 35°C and 36.5°C.

Wound temperature has influence on other parameters concerning
healing, in particular the rate of epidermal cell mitosis is
reduced if skin temperature falls[9]. Phagocytosis, by polymorpho-
nuclear leucocytes, which itself may help to prevent bacterial
growth, is increased at 37°C. These factors must be assessed when
deciding on an appropriate wound dressing.

    In the case of a surgical wound where little serous exudate is
expected but a high incidence of infection by commensal bacteria is
likely, a semi-permeable adhesive film dressing will ensure lower
skin surface temperatures leading to a reduction in septic complications.

REFERENCES

1. A.V.Pollock, Laparotomy, Proc. Roy. Soc. Med. 74: 480, (1981)
2. J.L. Ingraham, Growth of psychrophilic bacteria, J. Bacteriol.
        76: 75, (1958).
3. J.R. MacFarlane and M.L. Blowne, Cholecystectomy wounds: source
        of infection, J. Hosp. Infection, 3: 49, (1982).

4. G.D. Winter, Formation of the scab and rate of epithelialisation
    of superficial wounds in the skin of the young domestic
    pig, Nature, 193: 293, (1962).
5. C.D. Hinman and H. Harbach, Effect of air exposure and occlusion
    on experimental human skin wounds, Nature, 200: 377 (1962).
6. G.D. Winter, Oxygen and epidermal wound healing, Adv. Exp. Med.
    Biol. 94: 673, (1977).
7. T.K. Hunt, J. Niinikoski and B. Zederfeldt, Role of oxygen in
    repair processes, Acta. Chir. Scand. 138: 109 (1972).
8. I.L. Craft and H. Ellis, A method of evaluating surgical dress-
    ings, J. Clin. Res, 1: 5, (1968).
9. P.M. Lock, The effects of temperature on mitotic activity at
    the edge of experimental wounds, in "Wound Healing",
    praeparater, Oslo (1980).

CELLULAR RESPONSES TO SUTURES

Ian Capperauld

Research Unit
Ethicon Limited
Edinburgh, Scotland

INTRODUCTION

Suture materials are by far the commonest implant made into man. They are implanted into every site and every tissue of the body by the various surgical specialists. Sutures are placed in the heart and the brain; the lungs and pancreas; the stomach and colon; muscle and skin; blood vessels and the eye. A study of the differing cellular reactions to the various suture materials, therefore, should give a fairly extensive overview to the response of cells in general to all implants. Frequently, different types of suture materials are used during the same operative procedure and the implants are always multiple. To put the situation into perspective, approximately 9,000 operative procedures a day are conducted in the U.K., which will use approximately 60,000 sutures and ligatures. Certain sites in the body, especially the skin and the eye, allow direct clinical observation of the in situ reaction of these materials. The increasing use of endoscopes both in the gastro-intestinal and in the urinary tract also allows direct observation of the gross reaction to suture implants.

It should not be forgotten when considering the cellular reaction to sutures that there are two components occurring simultaneously. Firstly, there is the effect of the suture on the tissues and secondly, the effect of the tissues on the suture. Classically, the effect of the tissues on a suture is demonstrated when an absorbable suture, such as Catgut, is used and that material absorbs and disappears with time. This reaction is cellular and is necessary to allow the lysozymes carried within the polymorpho-nuclear leucocyte to be brought to the material to effect proteolysis. The suture material, if it were irritant or toxic, would also produce a cellular reaction.

Historically, the Edwin Smith Papyrus written 4,000 years ago, the Caraka Samhita written 3,000 years ago and the Shusrata Samhita written 2,500 years ago, all describe methods of closing surgical wounds by the use of a needle and some form of thread. The materials described there have been made from plants, fowls, ores and animals. They include cotton and linen from plants; tendons from chickens, falcons, kangaroos, Jack rabbits and oxen; silk from silkworms; and silver, gold and steel made from ores dug from the soil. Today, as well as synthetic polymers, even the shells of prawns and lobsters in the form of chitin, and the energy stores of bacteria in the form of polyhydroxybutyric acid, have been used to make sutures. In the surgical world today over 60% of all the suture material used probably still consists of the two oldest materials, namely silk from the silkworm and catgut made from the intestine of sheep or cow. Human tissue has also been used as sutures. McArthur in 1870 and Gallie in 1920 both stripped fascia lata from their patients' thighs to get suture material to repair their hernias. Mair in 1918 used an autogenous human skin strip to perform the same task. However, in the 1950's man's ingenuity through the good offices of the polymer chemist started to produce synthetic materials or man made fibres for use as sutures in surgery. Most of these fibres were non-absorbable but some of these new polymers were absorbable and hence a new generation of materials was born and a new challenge to the body's cellular reaction created.

CLASSIFICATION OF SUTURES

This appears complex, but an understanding of the different types of suture material is required to appreciate the various reactions which occur to them.

Absorbable

Catgut   Biological origin from sheep or ox; twisted, monofilament; undyed and uncoated; absorbable in approximately 90 days.

Collagen   Biological origin from ox; rolled monofilament; undyed and uncoated; absorbable in approximately 90 days.

Glycolide   Man made homopolymer; braided multifilament; dyed or undyed; coated or uncoated. Trade name DEXON (PGA - polyglycolic acid); absorbable in approximately 90 days.

| Glycolide & lactide | Man made copolymer; braided multifilament; dyed or undyed; coated or uncoated. Trade name VICRYL. Ratio of glycolide to lactide is 90/10 (polyglactin 910). Absorbable in approximately 90 days. |
| Polydioxanone | Man made copolymer; monofilament; dyed or undyed. Trade name PDS (polydioxanone suture). Absorbable in approximately 180 days. |

Non absorbable

Silk	Biological origin from silkworm; braided multifilament; dyed or undyed; coated or uncoated.
Linen	Biological origin from flax plant; twisted multifilament; dyed or undyed.
Cotton	Biological origin from cotton seed plant; twisted multifilament; dyed or undyed.
Polyester	Man made; multifilament; dyed or undyed; coated or uncoated.
Polyamide	Man made; monofilament or multifilament; dyed or undyed; generic name Nylon 6 or Nylon 66.
Polypropylene	Man made; monofilament; dyed or undyed.
Steel	Man made; monofilament or multifilament.

Sutures are attached to needles to allow tissues to be joined together while ligatures are lengths of suture materials used to tie off bleeding vessels and stumps of tissue.

The assessment of the cellular response to sutures can be performed in three ways:

1. In vivo implantation into the rat or rabbit.

2. In vivo using mouse fibroblast cells in tissue culture.

3. In situ in a clinical situation.

When examining a cellular response to a suture three reactions may be going on simultaneously. I believe it is important to separate these out when analysing the degree of tissue reaction and exactly what it means in terms of biological acceptability of the material by its host - Capperauld,[1].

The three reactions are:

1.   Reaction of implantation.

2.   Reaction of reaction.

3.   Reaction of absorption.

The reaction of implantation is self-explanatory. The degree of reaction can be influenced by the experience of the operator in handling tissues and the relative size of the suture to the animal's tissue. Reaction of reaction sounds a bit strange, but is a combination of factors. These include reaction to the shape of the suture, to the site of implantation, chemical composition of the material and, finally, whether the material is braided or mono-filament. I will show these in detail later. Lastly, the reaction of absorption applies mainly to absorbable materials and is a physiological mechanism to get rid of a foreign body. In general terms, the faster a material absorbs the greater the tissue reaction. Materials absorb either by enzymatic digestion, or by hydrolysis. Catgut and collagen absorb by proteolysis while the synthetic materials such as glycolide, lactide and dioxanone, absorb by hydrolysis.

The cellular response to suture materials is a dynamic, ever changing picture and pattern with time, so that any histological examination performed reveals a frame frozen picture only. Four features can be observed and measured: (i) size of reaction, (ii) type of cellular response, (iii) histochemical reaction, and (iv) longevity of the reaction.

Size of reaction

The size of the reaction is dependent on the site of implantation, the degree of trauma involved in implantation, the chemistry of the implant, the time after implantation the reaction is measured and the size and shape of the implant. In general, the more vascular the tissue the greater the reaction and of course this is compounded by the trauma of implantation. When the implant is of biological origin such as catgut, collagen, linen, cotton or silk, the reaction is much greater in size than when a synthetic polymer is implanted. The response to biological material is cellular and consists mainly of polymorphonuclear leucocytes and macrophages with occasional giant cells present (Figs. 1 and 2). Measurement of the size of reaction can be performed using the Sewell,[2] technique or by the use of a MOP or Manual Optical Processor. In the Sewell technique not only are the cells counted but they are also classified into various types and each type is assigned a weighting. The total reaction is, therefore, measured

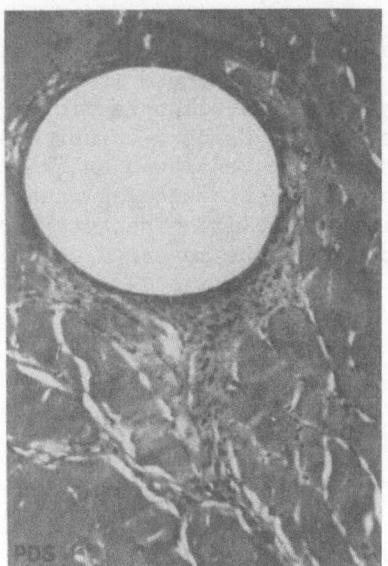
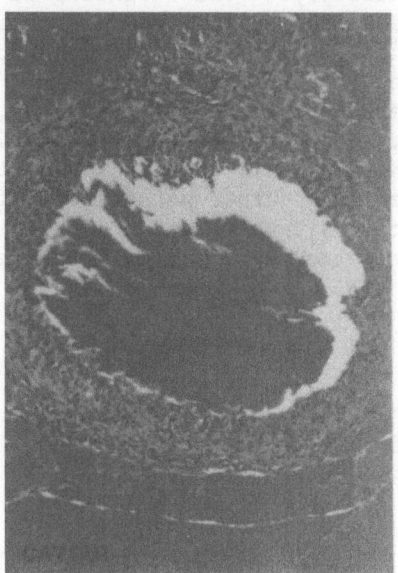

Fig. 1.   Photomicrograph of two monofilaments, PDS and Catgut, showing a much greater reaction to the biological suture material.

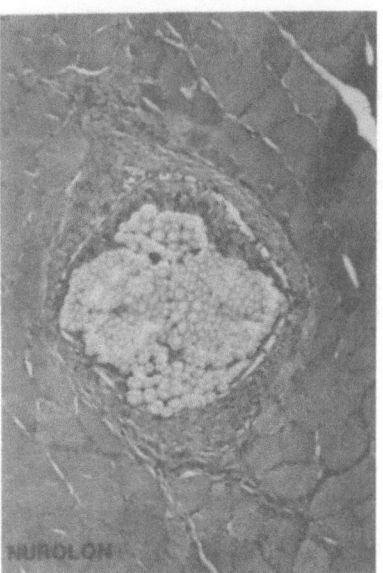
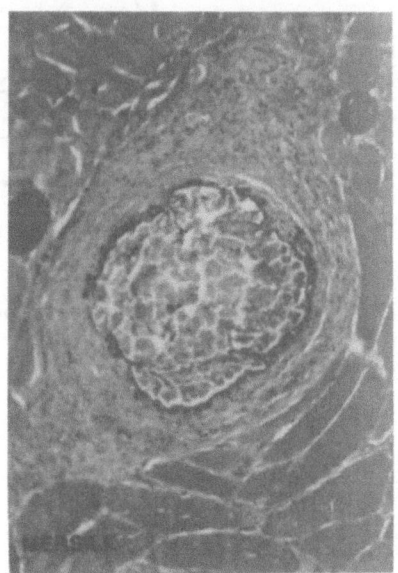

Fig. 2.   Photomicrograph of two braided sutures, NUROLON (braided nylon) and MERSILK, showing a much greater reaction to the biological suture.

by multiplying the weighting of each cell type with the numbers
present and adding them together. Using the MOP, however, is
easier to perform and is more accurate especially for comparative
assessment of serial sections. The instrument we use is a MOP-
Videoplan Image Analysis System, based on a 64-kilobyte micro-
processor with integral video display unit. Using a cursor and a
magnetised grid, superimposed structures can be viewed on the VDU,
allowing accurate measurement of many parameters such as area,
diameter, length, and counts of individual cells. The total area
of reaction surrounding a suture implant can be measured and

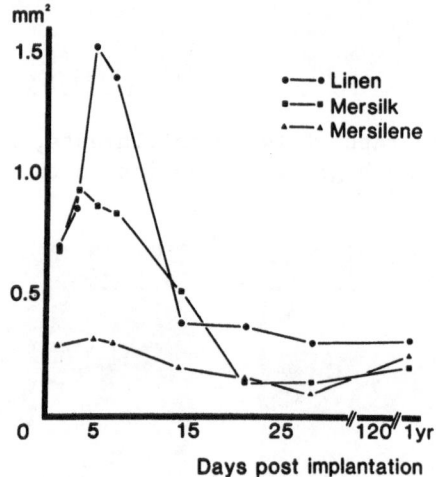

Fig. 3.   In vivo tissue reaction areas of Linen, MERSILK, and
          MERSILENE (braided polyester) measured on the MOP.

recorded in this way. Figs. 3, 4 and 5 show an experimental study
of seven non absorbable suture materials whose tissue reaction
was measured over a one year period. It is interesting to note
that synthetic material exhibits a lower take-off point than does
biological material, and the rise in reaction is not so great as
that for biological material.

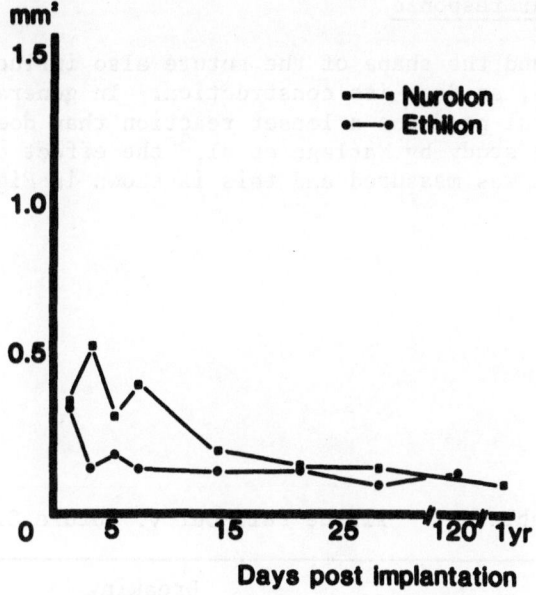

Fig. 4.   In vivo tissue reaction areas of NUROLON (braided nylon)
          and ETHILON (monofilament nylon) on the MOP.

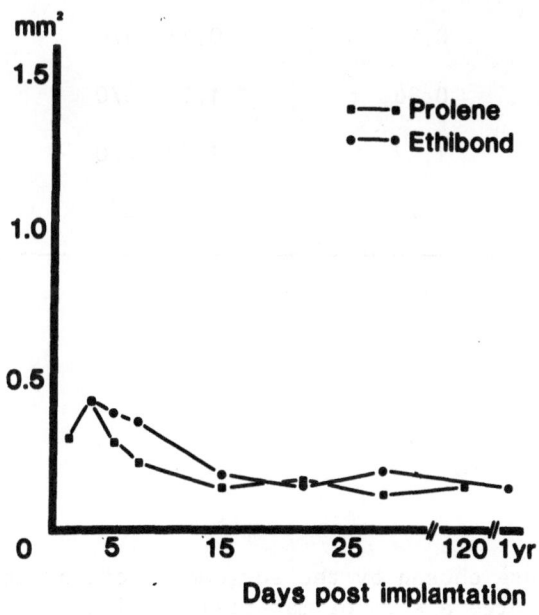

Fig. 5.   In vivo tissue reaction areas of ETHIBOND (polybutylate
          coated polyester) and PROLENE (polypropylene) measured on
          the MOP.

Type of cellular response

The size and the shape of the suture also influences the tissue reaction, as does its construction. In general, a mono-filament material produces a lesser reaction than does a braided material. In a study by Matlaga et al,[3] the effect of shape on tissue reaction was measured and this is shown in Figs. 6 and 7.

Table A - Tissue Pull-Out v. Suture Size

Tissue	Suture Pull-Out Value (kg)	Breaking Strength (kg) & Size of Catgut		Breaking Strength (kg) & Size of Silk	
Fat	0.2	0.31	6/0	0.20	7/0
Peritoneum	0.86	1.5	5/0	0.82	5/0
Muscle	1.27	1.70	4/0	1.70	4/0
Fascia	3.77	3.70	2/0	3.70	2/0

The size of suture chosen by the surgeon is often empirical and frequently too large a size of material is used. A study of the pull-out value of various tissues performed by Howes in 1929,[4] can be compared with a comparable tensile strength of the suture to derive the ideal size to be used in a particular procedure and is shown in Table A, Capperauld,[5].

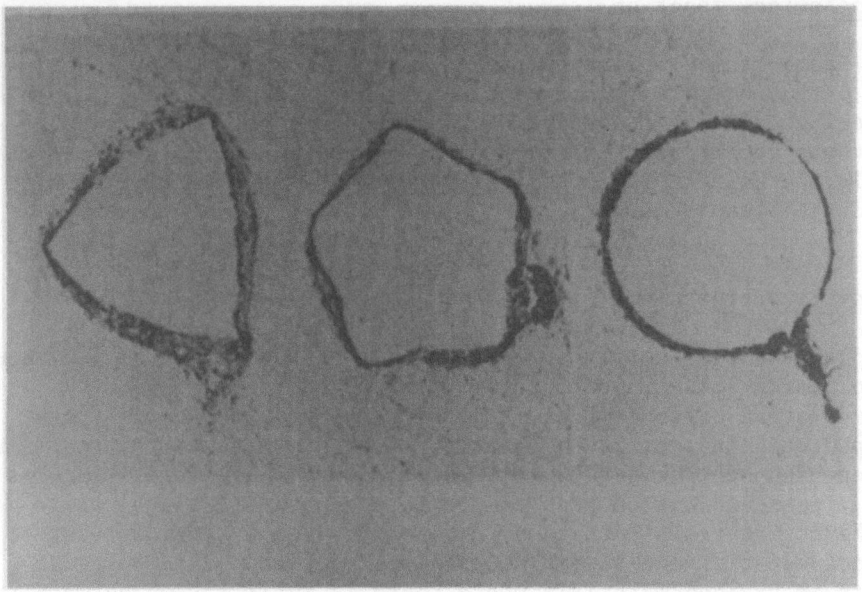

Fig. 6.   Enzyme activity around three configurations of P.V.C.
          implants at 14 days.

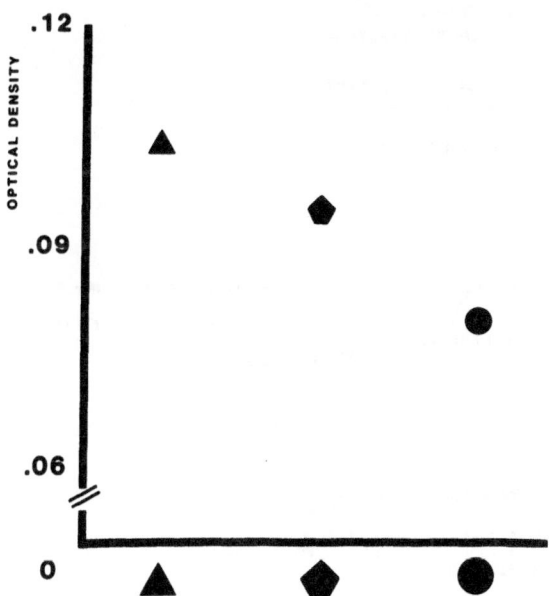

Fig. 7.   Average enzyme activity readings of the three implanted
          P.V.C. shapes.

Histochemical reaction

        Histochemical reactions to sutures have been measured, mainly
with catgut where the acid phosphatase and aminopeptidase levels at
differing periods of absorption have been measured, Salthouse et al,[6]
and the typical graph of reaction is shown in Fig. 8. While these
levels are related to tensile strength loss, they in turn also
reflect tissue reactivity.

Longevity of reaction

        The end result of tissue reaction to a suture material causes
one of three situations to occur, that of encapsulation,
incorporation or absorption. Most non-absorbable materials are
encapsulated in a collagen capsule. The thickness of this capsule
is dependent on the suture material. Biological material such as
silk, linen and cotton result in a thick capsule (Fig. 9) while
synthetic materials such as nylon, polypropylene, stainless steel
and polyester give a very fine collagen capsule. Incorporation
occurs in braided materials where there is infiltration between the
fibrils of the braid by collagen fibres and the material becomes an
inherent part of the tissues requiring quite an effort to separate
the suture from its host.

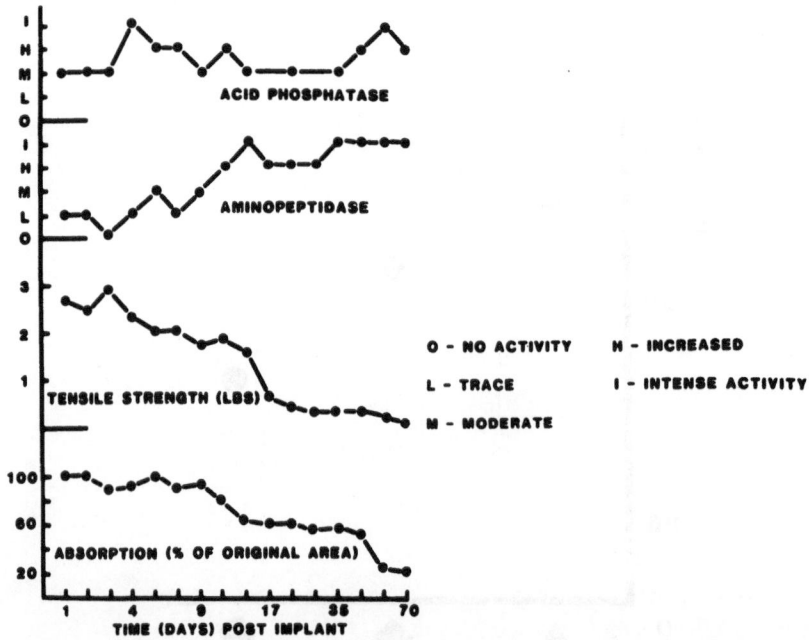

Fig. 8. A comparison of tensile strength, absorption and histo-
         chemical activity of implanted collagen suture over a
         given time period.

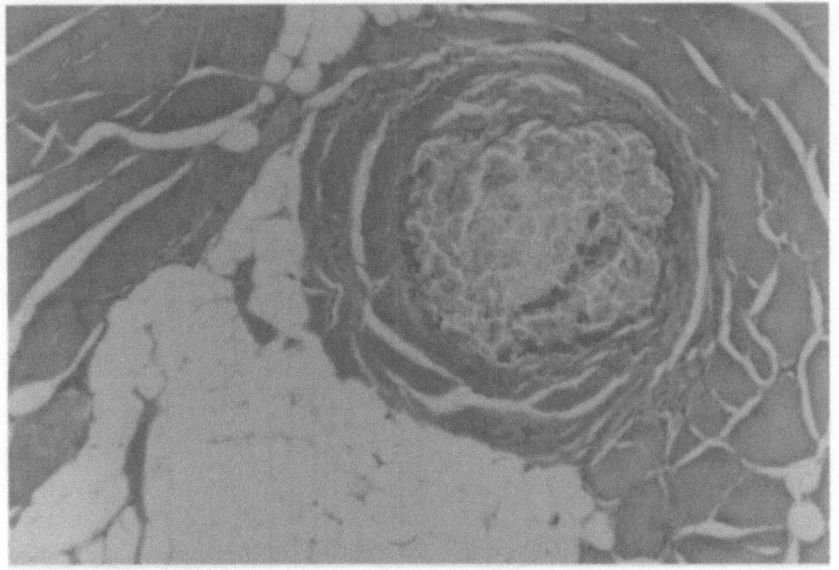

Fig. 9. A section of MERSILK at one year post implantation showing
complete encapsulation by many layers of collagen.

Finally, absorption is shown in catgut and collagen by a
progressive decrease in size of the suture with fragmentation and
separation of the ply bonding until finally a small collagen scar
is left. With synthetic absorbables such as DEXON (homopolymer of
glycolic acid), VICRYL (copolymer of glycolide and lactide) and PDS
(homopolymer of dioxanone) there is progressive loss of crystallinity
with change in histological staining reaction and finally a decrease
in volume of the material until finally a small collagen scar is
left. With catgut and collagen this takes approximately 60 - 90
days depending on whether the material has been chromicised or is
plain. With DEXON and VICRYL this takes 90 ± 30 days while PDS
takes approximately 180 days to disappear and be replaced with a
fine scar.

TISSUE CULTURE

The cellular response to a suture material can be measured on
tissue culture using mouse fibroblasts and an overlay technique as
shown in Fig. 10. By using this technique the biocompatibility from
an early stage can be demonstrated. The greatest 'toxic' reaction
is shown to silk, while the least 'toxic' reaction is shown to
polypropylene - Figs. 11 and 12.

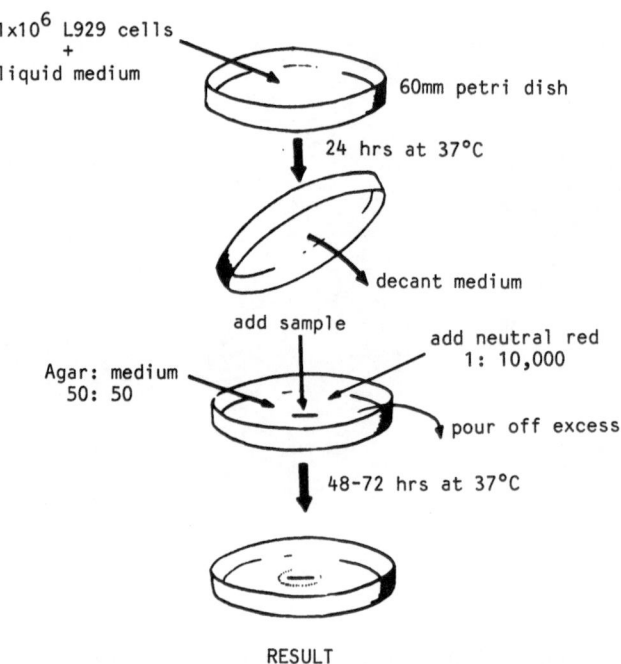

Fig. 10. Diagrammatic outline of the agar overlay cytotoxicity
          technique.

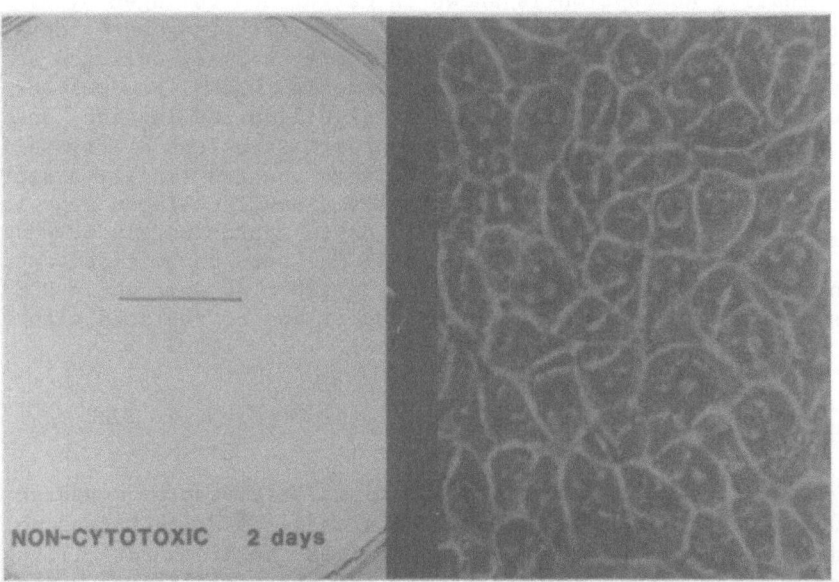

Fig. 11. Zone of colour loss comprising dead mouse fibroblasts
          indicating 'toxic' reaction to MERSILK.

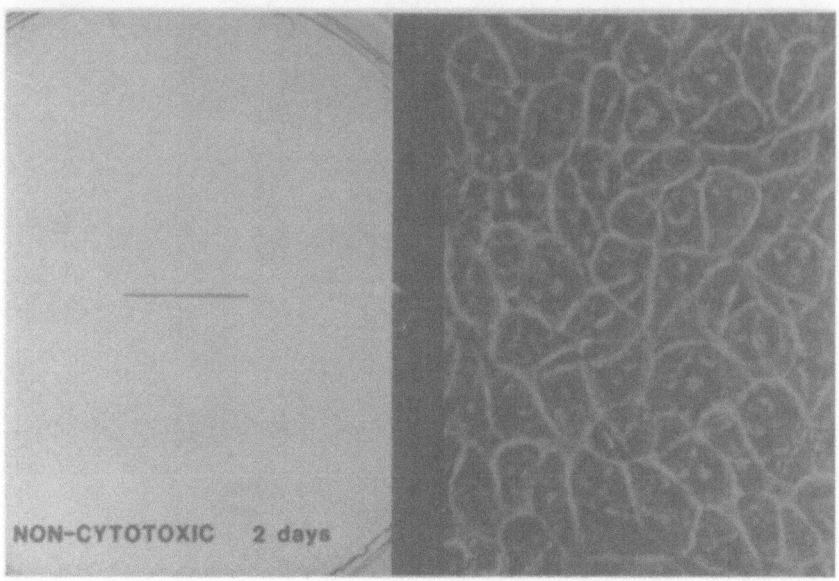

Fig. 12.   No zone develops with polypropylene suture.   Cells remain
           stained and viable.

Complete biocompatibility is shown to most synthetics, but
polypropylene in particular yields a good result.   Scanning
electron microscopy can be performed on sutures retrieved from
tissue culture and this shows total acceptability to the material -
Figs. 13 and 14.   There is no change in multiplication of the
fibroblasts nor is there any cellular death with materials like
polypropylene and polyester.

CLINICAL OBSERVATION

The cellular response to suture material can be observed in
the skin when sutures are left in for varying periods to hold the
wound together.   In operations involving the eye, suture materials
again can be directly observed and even more critically examined
using an ophthalmoscope or using a slit lamp.

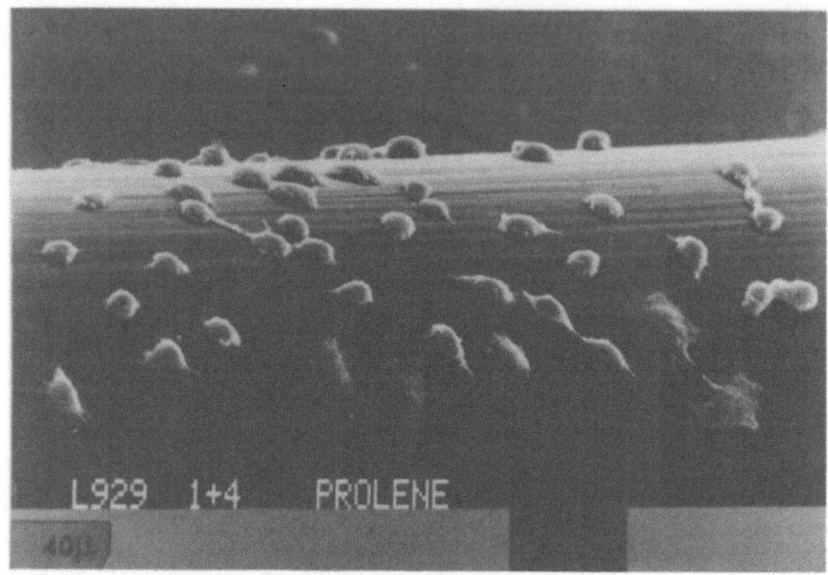

Fig.   13.   An in vitro study, showing scant mouse fibroblasts
             attached to a polypropylene suture, indicating complete
             biocompatibility.

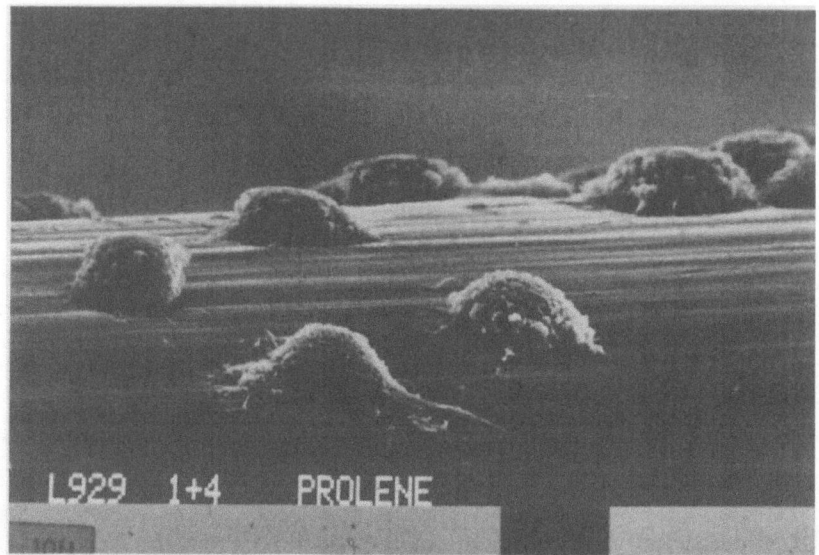

Fig. 14.   Attachment of the individual mouse fibroblasts to a
           polypropylene suture in vitro.

CONCLUSIONS

     Suture materials are by far the commonest implant known to man.
In the U.K. in excess of 500,000+ such implants are made weekly into
man during the course of 55,000 surgical operations. A study of
the reactions to these materials gives a very wide spectrum of the
reactions to various biological and synthetic materials, and gives
a very good insight into tissue reactions in general to all implants.

REFERENCES

1.   I. Capperauld, Bovine fibrin implants: tissue reaction, in:
     "Biomaterials in Reconstructive Surgery," L. R. Rubin, ed.,
     C. V. Mosby, St. Louis (1983).
2.   W. R. Sewell, J. Wiland, and B. N. Craver, A New Method of
     Comparing Sutures of Ovine Catgut with Sutures of Bovine
     Catgut in Three Species, Surg. Gyn. Obst. 100:483 (1955).
3.   B. F. Matlaga, L. P. Yasenchak, and T. N. Salthouse, Tissue
     Response to Implanted Polymers: The Significance of Sample
     Shape, J. Biomed. Mater. Res. 10:391 (1976).
4.   E. L. Howes, and S. C. Harvey, The strength of healing wound
     in relation to the holding strength of the catgut suture,
     New Eng. J. Med. 200:1285 (1929).
5.   I. Capperauld, and T. E. Bucknall, Sutures and Dressings, in:
     "Wound Healing for Surgeons," T. E. Bucknall and H. Ellis,
     ed., Bailliere Tindall, London (1984).
6.   T. N. Salthouse, J. A. Williams, and D. A. Willigan,
     Relationship of Cellular Enzyme Activity to Catgut and
     Collagen Suture Absorption, Surg. Gyn. Obst., 129:690 (1969).

SECTION 6

OTHER CLINICAL ASPECTS OF BIOCOMPATIBILITY

AN EXPERIMENTAL STUDY OF CALCULOGENESIS AND BIOCOMPATIBILITY

OF PLASTICS FOR UROLOGICAL USE

P.R. Crocker, J.W.A. Ramsay, S.F. Hill and H.N. Whitfield

Departments of Histopathology and Urology
St. Bartholomew's Hospital
West Smithfield
London EC1

SUMMARY

Silicone, Polyvinylchloride, Nylon and a Polyethylene Copolymer were studied in the urinary tracts of experimental animals. Continuous administration of Ascorbic Acid Sulphadimidine was used to control urinary pH and infection. Scanning Electron Microscopy with a Back Scattered mode was used to examine the surface of the tubes and to define the inorganic element composition of the encrustations, by X-ray Energy Spectroscopy. To check the compound composition of the encrustations Infra-red Spectroscopy was performed. Light microscopy was used to compare the urothelial surfaces in contact with the implants.

INTRODUCTION

Preliminary reports of the in vivo performance of acrylic acid graft copolymers suggested a high incidence of encrustation and stone formation when implanted into the urinary bladders of Wistar rats. Acrylic acid graft copolymers which swell but do not dissolve in water have a recognised potential as biomaterials. It was therefore important to compare a Polyethylene copolymer with established biomaterials in controlled experimental conditions. Scanning electron microscopy has been employed to compare the surface properties of urethral catheters which have been found to be cytotoxic to cell cultures as the conclusion that surface morphology did not relate to the cytotoxicity of these catheters was clinically important. However, the principal clinical problem which frustrates the use of much smaller calibre ureteric stents is blockage and encrustation. This study was designed to investigate stent surface morphology and composition and to relate this to the degree of encrustation under controlled experimental conditions.

TABLE 1. EXPERIMENTAL GROUPS

Composition of tube	Additive	Number of animals implanted
Silicone	Barium sulphate	3
Polyvinylchloride	Barium sulphate	3
Polyvinylchloride	Barium chloride	3
Nylon	-	3
Polyethylene copolymer	Polyvinyl alcohol coating	3

MATERIALS AND METHODS

1.   Animals: 15 female New Zealand White (NZW) rabbits were used.
Anaesthesia was induced by intramuscular Hypnorm(Fentanyl and
Fluanisone) 0.4ml kg and maintained by 0.1 ml intravenous aliquots
of the same drug. Relaxation was enhanced by the local infiltration of 1% lignocaine solution. 2 - 4 FG (overall diameter o.7 mm
- 1.4 mm)tubes were introduced into the left ureter through a mid-
line vesicotomy. Five types of ureteric stent (Table 1) remained
indwelling for 1 month. Each animal received 125 mg Ampicillin
intramuscularly prior to operation and 500 mg Sulphadimidine per
day throughout the experiment. In addition 2g Ascorbic Acid were
added to 300 mls of the drinking water for each 24 hr period. One
month after implantation the left kidney and ureter were removed
en bloc with the bladder; the plastic tube was removed from the
ureter, and in the case of the hydrophillic copolymer preserved in
the rabbit urine until analysis. A section of ureter was fixed in
a solution of 4% formaldehyde in 0.9% sodium chloride for light
microscopy. Urine cultures were obtained peroperatively at the
beginning and end of the study by needle aspiration. pH measurements were made on the urine specimens from days 1 and 28, and in
6 animals on day 14.

2.   Electron microscopy and stone analysis: Samples of clean and
experimentally used tubes were prepared in an identical manner for
examination by scanning electron microscopy. For longitudinal
examination 2 mm pieces, and for cross section examination 0.5 mm
pieces were cut. These were then mounted on to perspex 3 x 1 inch
slides by means of double sided sellotape, carbon coated in a
Nanotech CC2 carbon coater, and placed in a JOEL 35CF scanning
electron microscope fitted with a backscattered electron detector,

TABLE 2. INCIDENCE OF ENCRUSTATION ON URINARY TRACT IMPLANTS
REMOVED AT 1 MONTH

Animal	Implant	Urinary pH	Analysis	
			XES	IR
29	Silicone	8.0	Ca, Ba	$CaCo_3$
38	(Ba)	7.5	Ca, Ba	$CaCo_3$
42		7.0	Ba	–
4	PVC (Ba)	8.0	Ca, Ba	$CaCo_3$
39		8.5	Ca, Ba	$CaCo_3$
41		8.0	Ca, Ba	$CaCo_3$
34	PVC (Bi)	8.0	Ca, Bi	$CaCo_3$
49		7.5	Ca, Bi	$CaCo_3$
50		8.0	Ca, Bi	$CaCo_3$
35	Nylon	7.5	Ca	$CaCo_3$
36		8.0	Ca	$CaCo_3$
37		8.5	Ca,Mg	$CaCo_3$
19	Polyethylene	8.0	K+	–
45	copolymer	7.5	Ca, K+	$CaCo_3$
44		7.5	Al,Ca, K+	$CaCo_3$

a Kevex Unispec System 7000 X-ray energy spectrometer, and a Dapple
Microplus data processor. The surfaces were then examined in the
scanning electron mode (SE1) and by the backscattered mode (BE1)
at 25kv. Areas of atomic number difference revealed by the BE1 mode
from particulate matter on the surface or within the lumen of the
tubes were then analysed for their inorganic element content by
X-ray energy spectroscopy (XES).

To study the surface detail of the tubes further samples were
prepared as above and gold coated in a Emscope SC500 gold coater
and examined by scanning electron microscopy at 15 kv.

Where sufficient particulate material was present in the experimental tubes compound analysis was performed by Infra-red
Spectroscopy (IR) to determine its chemical nature. The samples

were examined by compressing 1.5 mgms of the sample with 300 mgms
of potassium bromide and producing a 13 mm disk for examination in
a Perkin Elmer 577 infra-red spectrophotometer.

RESULTS AND DISCUSSION

The incidence of encrustation is shown in Table 2. All animals
remained uninfected. A control group of unoperated animals had an
average urinary pH 9.0. Urinary pH averaged 8.0 in this experiment.
Minor urothelial abnormalities were encountered in 3 animals implanted with nylon, PVC(Bi) and Polyethylene copolymer (Figure 1).
All stented ureters were dilated, but only two kidneys were hydro-
nephrotic. Figure 2 shows SEM and XES of an encrustation from a
bismuth containing stent. The bismuth occurs within the encrustation.

Rabbit urine produces a favourable medium for encrustation
with $CaCo_3$ onto all implanted materials. By reducing urinary pH
below the point of precipitation of these salts less encrustation
might be predicted. In practical terms the urinary pH could not
be reduced to this level. The rabbit urinary tract, therefore,
provides a rigorous model for calculogenesis onto biomaterials.
As it was possible to control urinary infection, any differences
in degree of calculogenesis may be related to the surface morphology
of the biomaterial. In this respect Polyethylene Copolymers and
Silicone compared favourably. The surface morphology of wetted
Copolymers is similar to Silicone (Figure 3). Bismuth is added
to polymers to confer radiopacity. The appearance of bismuth in
the encrustations around such biomaterials has not been previously
reported. Little significance should be attached to this isolated
finding in an experimental model. However, it is likely that
certain polymers may be partially soluble under extremes of urinary
pH.

The finding of hydroureter without hydronephrosis in stented
ureters may be of clinical significance. The ureteric stents examined were relatively free of encrustations on their outer surfaces, except in the position distal to the ureterovesical junction
(Figure 4).

The lumena of the stents were blocked in all cases. The
ability of urine to drain around a patent ureteric stent, has been
previously reported [4]. These observations highlight the importance
of urinary flow over a biomaterial to reduce the rate and degree
of encrustation. The advantage of hydrophillic plastics in this
particular application may be reduced internal encrustation due to
increased intraluminal diameter and flow.

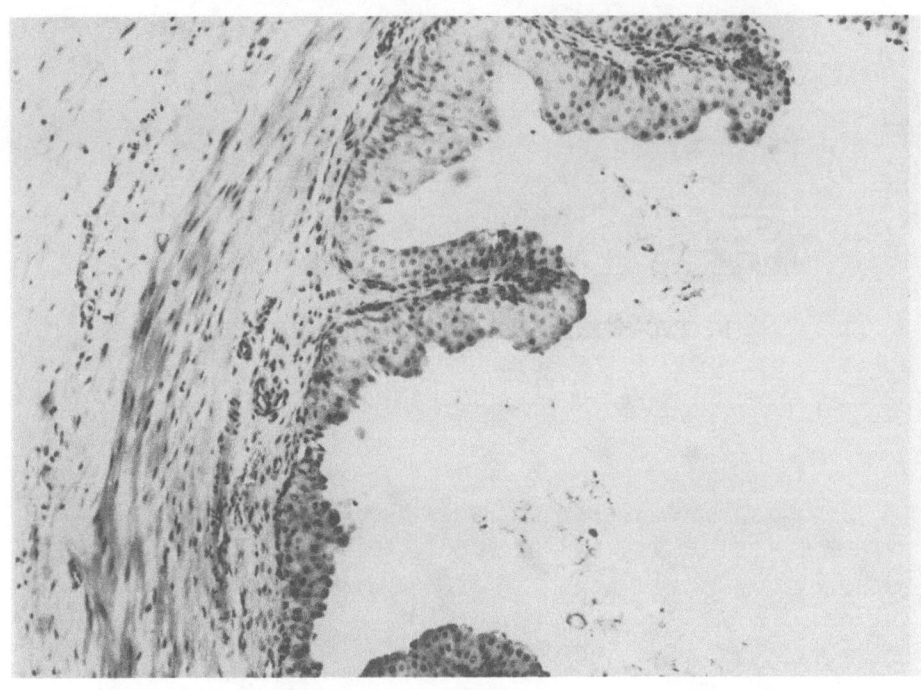

FIG. 1. UROTHELIAL ABNORMALITIES   a) NYLON STENT
(Continued)

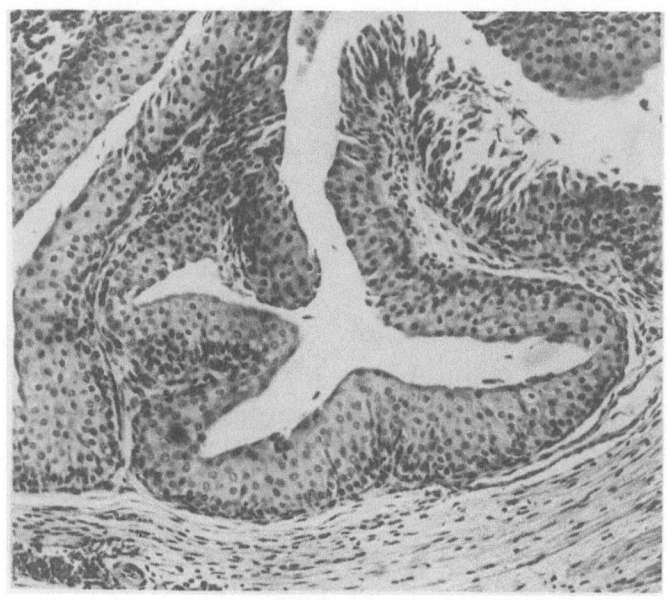

b. PVC STENT IMPREGNATED WITH BISMUTH

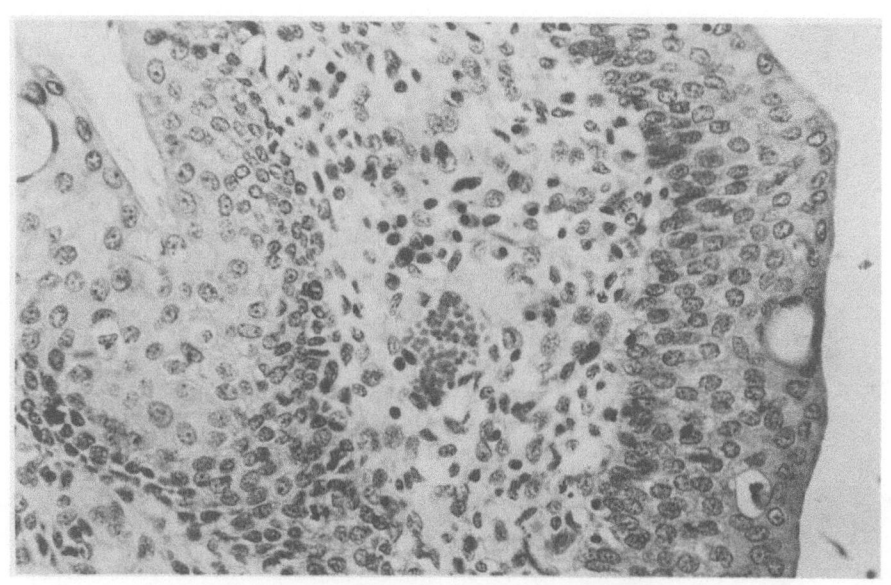

c. POLYETHYLENE COPOLYMER

FIG. 1.   UROTHELIAL ABNORMALITIES

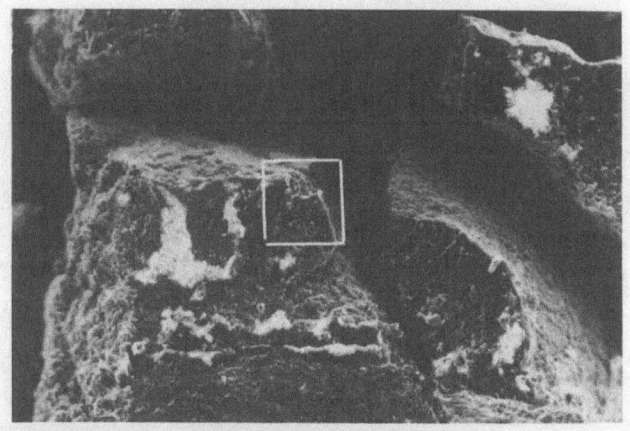

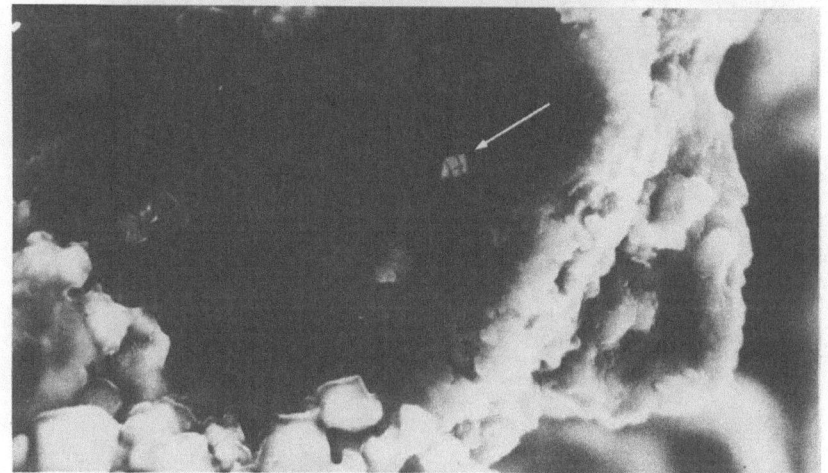

FIG. 2. SEM AND BEI MODES OF A ENCRUSTATION FROM A BISMUTH
IMPREGNATED PVC STENT

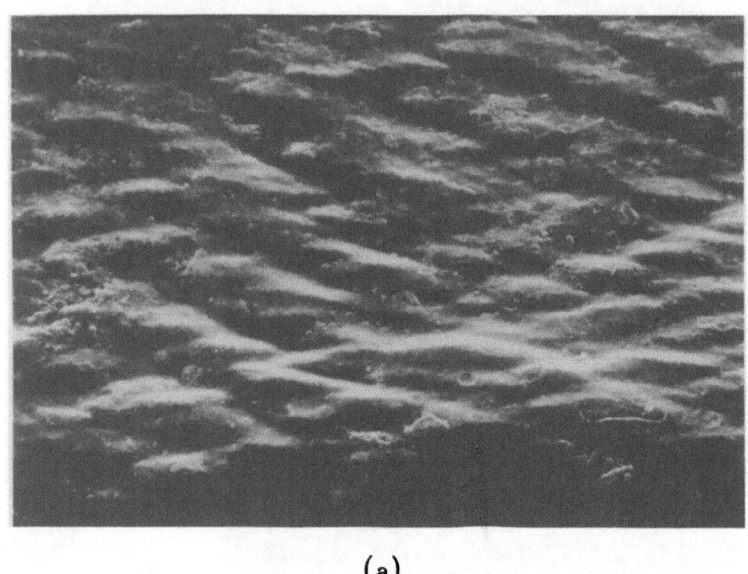

(a)

(b)

FIG. 3 THE SURFACE MORPHOLOGY OF (a) DRY POLYETHYLENE COPOLYMER,
(b) SILICONE, AND (c) WET POLYETHYLENE COPOLYMER

(c)

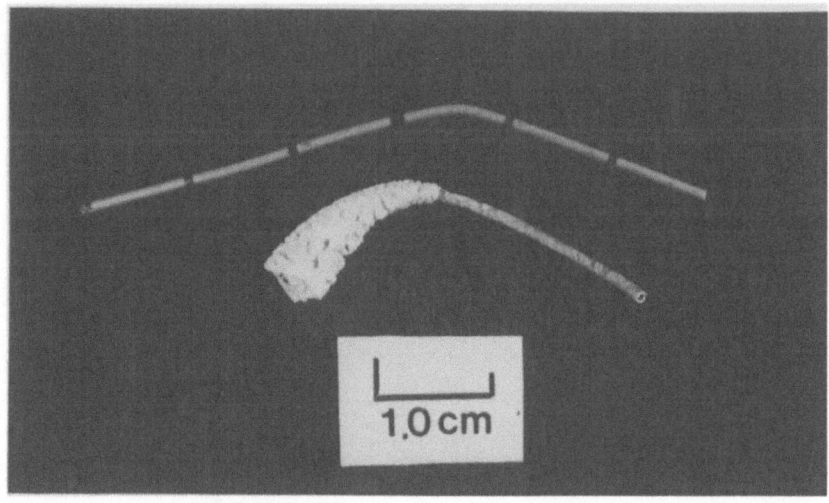

FIG. 4   A CLEAN AND ENCRUSTED URETERIC STENT TO SHOW INCREASE
IN ENCRUSTATION IN THE INTRAVESICAL PORTION

REFERENCES

1. T.F. Ford, M.C. Parkinson, P.J. Fydelor, B.J. Ringrose and
      J.E.A. Wickham. A preliminary in vivo assessment of Acrylic
      acid graft copolymers in the urinary tract. J. Urol. 133:
      1-3, (1985).
2. P.J. Fydelor and D.E.M. Taylor, Biocompatible surgical devices.
      U.K. Patent 2035350B, (1981).
3. J. Wilksch, B. Vernon-Roberts, R. Garrett and K. Smith, The
      role of catheter surface morphology and extractable cytotoxic
      material in tissue reactions to urethral catheters.
      British Journals of Urology 55: 48-52, (1983).
4. J.W.A. Ramsay and H.N. Whitfield, Bilateral pelviureteric junc-
      tion obstruction; the case for endoscopic management.
      Journals of the Royal Society of Medicine (1985) in press.

ACKNOWLEDGEMENTS

    The authors wish to thank Professor David Taylor for his
guidance and assistance in this project, the North East Thames
Regional Health Authority, the Wellcome Trust and the Joint Research
Board of St. Bartholomew's Hospital who together financed the purchase
of our analytical apparatus. The expert photographic assistance
provided by Mrs. Dawn O'Keafe and Mr. John Hopwood was greatly
appreciated.

THE MECHANICAL AND BIOLOGICAL PROPERTIES OF FILAMENTOUS CARBON OR POLYPROPYLENE MESH IN THE RAT ABDOMINAL WALL

Alan Cameron and David Taylor

Department of Applied Physiology & Surgical Sciences

Royal College of Surgeons of England

INTRODUCTION

Prosthetic repairs of tissue defects may fail at the junction with normal tissue because this is a region of high stress due to a fundamental mismatch in mechanical behaviour [1]. This mismatch may reflect poor biological function due to the poor ingrowth of host fibrous tissue.

Carbon fibre implants have been shown to act as a scaffold for well-organised collagen in tendons [2], so we have studied this material and polypropylene mesh in the repair of experimental abdominal wall defects in rats.

MATERIAL AND METHOD

a) Creation of Defects

Surgery was carried out on four groups of Porton-Wistar rats aged 6-9 months under pentobarbitone anaesthesia. In the first three groups a vertical skin incision was made for xiphisternum to pubis, and then the abdominal wall muscles were excised widely. The amount of tissue excised was weighed. The fourth group received the anaesthetic and the skin incision only.

b) Repair

In Group A the skin was closed with no deep repair. In Group B the defect was replaced by a polypropylene (Marlex) mesh, secured by continuous 6/0 polypropylene (Prolene). In Group C the defect was repaired by an open, simple, side-to-side reef of carbon fibre.

Carbon fibre (Grafil X-AS, Courtaulds Ltd.) 6000 filaments thick
and carried on a round-bodied needle, had been sterilised by gamma-
irradiation.

c)   Assessment of Results

     After five months the 'clinical' result was noted and the
animals sacrificed. The abdominal wall was excised, trimmed of skin
and fat, and a 1cm wide sagittal strip was cut from the centre of
the repair, extending laterally into normal tissue. The load
required to disrupt the tissue strip and its extension to breaking
point were recorded, using an Instron tensiometer (Model 1026)
applying a steady extension of 50mm min^{-1}.

RESULTS

a)   Clinical

     There were nine survivors in Group A (no repair), 11 each in
Group B (Marlex) and Group C (carbon fibre), and 10 in Group D (skin
incision only). The amount of tissue excised did not differ significantly between the three operated groups (Table 1). All of the unrepaired defects resulted in huge ventral hernias. There were no
gross hernias in either of the repaired groups, but in four of the
animals repaired with carbon there were one or more small bulges.
Both types of repair were well-tolerated macroscopically.

Table 1. Extent of abdominal wall resection, and tensiometric
          estimation of breaking stress of a 1cm strip of abdominal
          wall elastance after 5 months survival (Mean and range).

Group	A No repair	B Marlex	C Carbon	D 'Sham' Control
Survivors	9	11	11	10
Tissue excised (g)	3.2 (1.4-4.6)	2.9 (1.9-4.8)	2.9 (1.5-4.8)	0
Breaking load (N.cm^{-1})	7.8 (1.5-15.2)	19.2 (12.2-32.9)	17.2 (6.8-30.4)	20.5 (11.3-31.9)
Elastance (Nm^{-1})	2109 (834-4765)	3876 (1907-5935)	3817 (1552-7347)	3782 (2456-5130)

b)   Mechanical

        The results of tensiometric testing are summarised in Table 1.
The breaking loads showed that both the Marlex repair (p < 0.01)
and the carbon repair (p < 0.02) were significantly stronger than

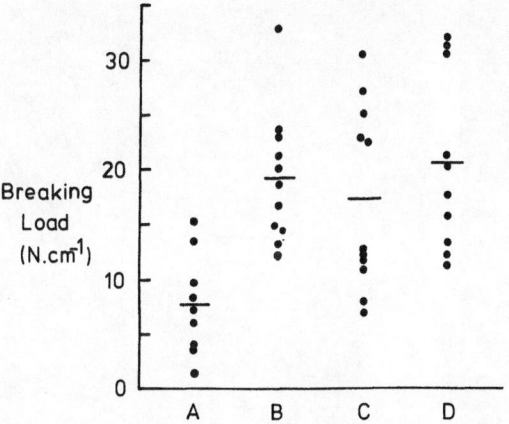

Fig. 1.   Breaking load of abdominal repair for 1cm strip, with mean
           indicated by bar.   (A - No repair: B - Marlex mesh: C -
           carbon fibre darn: D - sham operated).

the unrepaired (Fig.1).   There were no significant differences in
breaking load between the two repaired groups and the sham operated
animals.

        The elastance (force required to produce unit elongation) of
the four groups showed an identical pattern to that of breaking
load (Table 1).

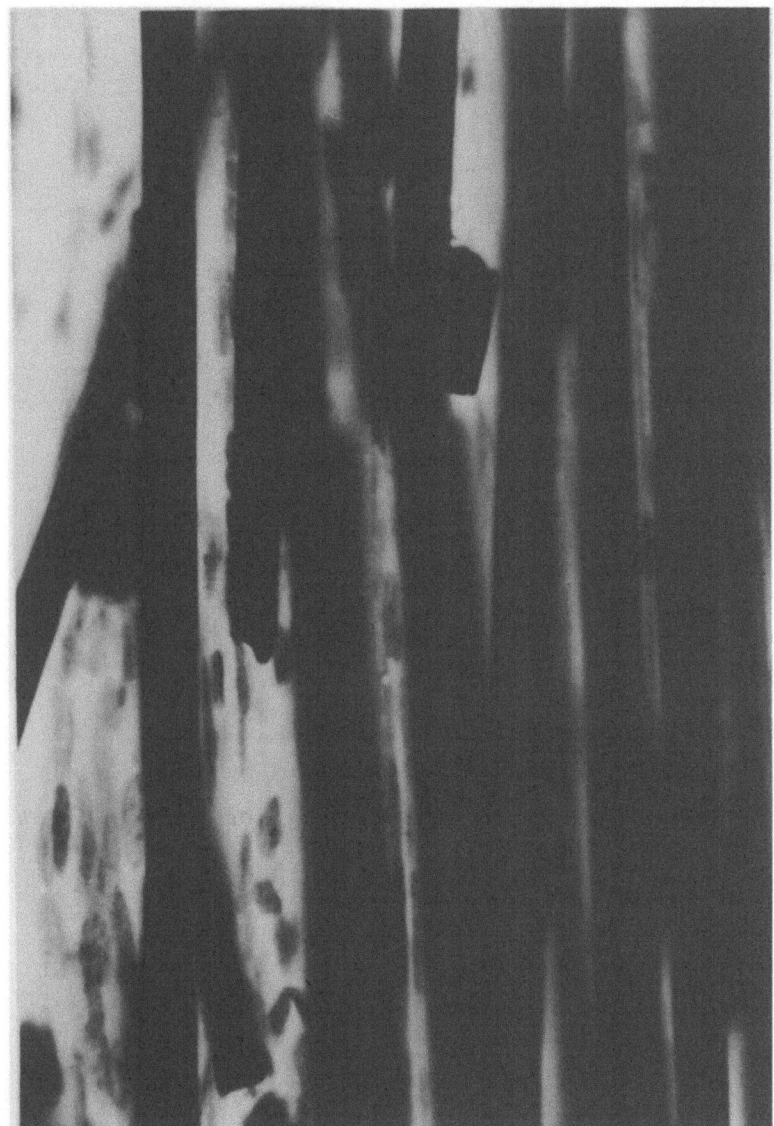

Fig. 2. Histological appearance of the abdominal wall repaired
         with carbon fibre. Note that the round cell infiltrate
         is very much less than with the Marlex repair (Fig.4).
         Further, the collagen is well aligned parallel to the
         carbon fibrils.

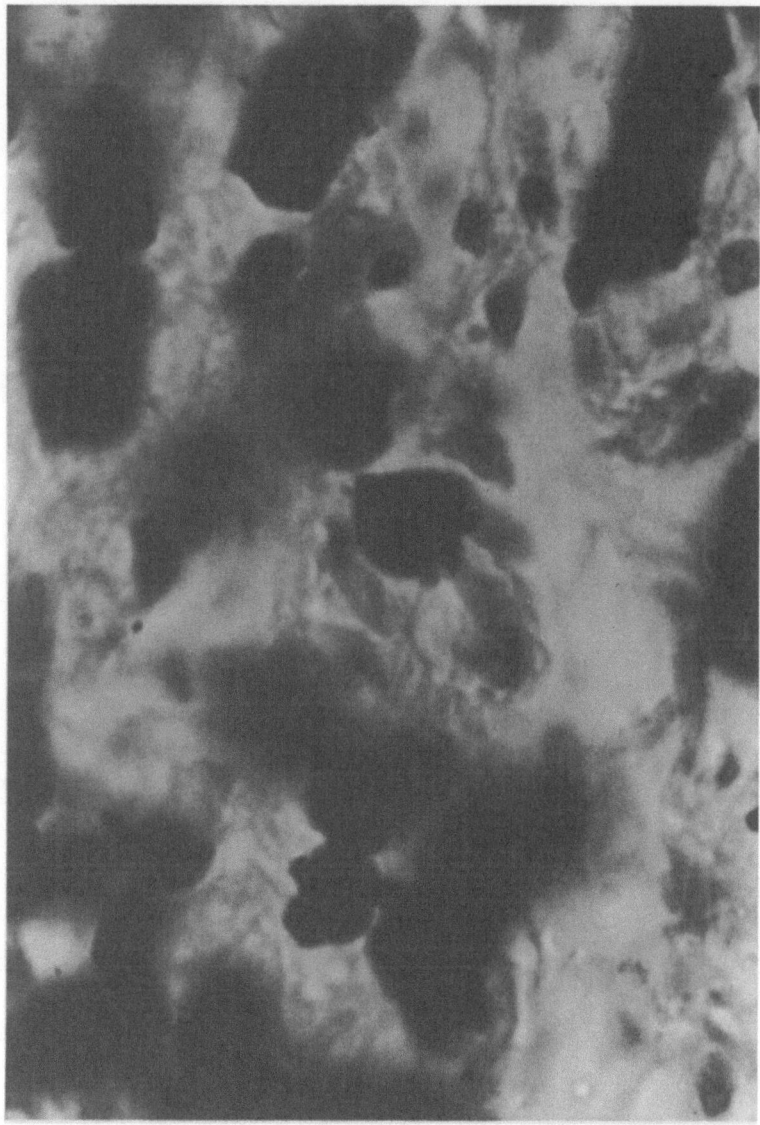

Fig. 3. Histological appearance of the abdominal wall repaired with carbon fibre. The section has been cut normal to the carbon fibrils. Note again the orientation of the collagen fibrils to be parallel with the carbon fibres and minimal round cell infiltrate.

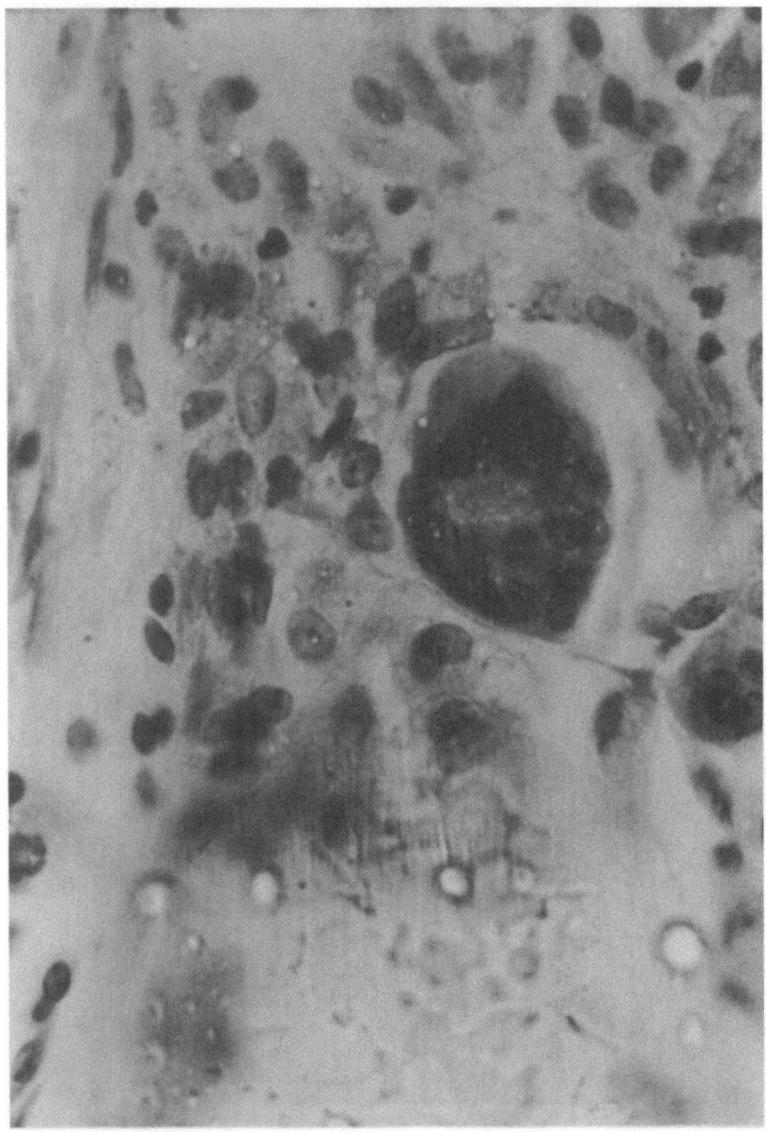

Fig. 4. Histological appearance of the abdominal wall repaired with
         polypropylene (Marlex) mesh. Note that there are numerous
         round cells and occasional giant cells. The collagen
         fibrials are not well aligned.

c)   Histological

The carbon fibre implants had been fully infiltrated by well-orientated collagen fibres. The host tissue was in alignment with the fibres and there were few round cells and few giant cells (Fig. 2). In cross section each carbon filament was surrounded by parallel collagen bundles (Fig.3). In contrast, the collagen related to the Marlex mesh was arranged in a circular, disorganised fashion. There was a marked infiltrate of large numbers of round cells and frequent giant cells, suggesting a chronic foreign-body type of inflammatory response (Fig.4).

DISCUSSION

Carbon fibre has been implanted in man [2,3] , sheep [4] and rabbits [1] to repair or replace damaged tendons and ligaments. In each case it has induced the ingrowth of collagenous tissue closely resembling the host tissues and this has been reflected in an excellent clinical outcome. The present study confirms that filamentous carbon behaves in the same way in the rat abdominal wall. Although there were no hernias in the carbon group, the bulging observed could represent a potential recurrence. However, this was due to inappropriate technique rather than material failure: the rats were extremely active immediately after surgery and it was unrealistic to expect that the viscera would be contained by such an open weave. The interface between the implant and the repair is the region maximally stressed during tensiometric testing, and the carbon repair was as strong as the Marlex repair in this study which indicates the excellent fixation induced by the relatively narrow carbon suture.

A previous qualitative study in sheep compared a side-to-side reef of carbon fibre with a polyester (Mersilene) mesh repair of the abdominal wall [5]. The recurrence rate was lower after the carbon repair, and there was a marked deposition of fibrous tissue in the interstices of the carbon but not in Mersilene. We chose to compare carbon with polypropylene rather than polyester because the latter can produce a marked tissue response. For example with vascular prostheses [6], but we did still observe chronic inflammatory cells around the polypropylene. The higher biotolerance of the carbon permitted tissue ingrowth throughout the interstices of the fibre. The relatively high surface area to mass, which can be a significant factor in the tissue response to less biotolerant materials [6], did not appear important with carbon. Carbon fibre is already in use in orthopaedic surgery as a tendon replacement, and it may have a place in general surgery for the repair of difficult hernias. Repair of such defects with currently-available mesh prostheses has not been totally satisfactory, possibly because of foreign-body reaction, lack of tissue ingrowth, and recurrence due to suture slippage [7].

Conventional repairs merely provide mechanical support; as a concomitant carbon fibre darn or carbon fibre textile insert might also induce endogenous supporting tissue, the material deserves further study.

ACKNOWLEDGEMENTS

We are grateful to Dr. Peter F. Lofts for stimulating our interest in carbon fibre and initiating this study, which was funded by Johnson & Johnson Ltd.

REFERENCES

1. J. Aragona, J.R. Parsons, J. Alexander and A.B. Weiss, Soft tissue attachment of a filamentous carbon-absorbable polymer tendon and ligament replacement, J. Bone and Joint Surg (Br), 61B: 120 (1979).
2. D.H.R. Jenkins and B. McKibbin, The role of flexible carbon-fibre implants as tendon and ligament substitute in clinical practice, J. Bone and Joint Surg(Br). 62B (4): 497-99 (1980).
3. Beverley Kramer and R.E. King, The histological appearance of carbon-fibre implants and neoligaments in man, S.A. Med. J. 63: 113-15, (1983).
4. D.H.R. Jenkins, The induction of new anterior cruciate ligaments in the sheep and the clinical use of filamentous carbon fibre in the human, J. Bone and Joint Surg(Br). 61B: 129 (1979).
5. C, Johnson-Nurse and D.H.R. Jenkins, The use of flexible carbon fibre in the repair of large abdominal incisional hernias. Br. J. Surg. 67: 135-37 (1980).
6. D.E.M. Taylor, J.S. Whamond and E.J. Santiago, Effect of method of construction on fibrous ingrowth into knitted polyester arterial prostheses, Proceedings IX Annual Meeting EASO, Brussels, 269-72, (1982).
7. R.J. Minns, D.C. Stevens, D.M. Gore and L.F. Tinckler, The mechanical and structural properties of reinforcing materials used in prosthetic herniorrhaphy, Eng. in Med. 8: 15-20, (1979).

THE POSSIBLE ROLE OF IMPLANT MATERIALS IN PROMOTING

THE ASEPTIC LOOSENING OF PROSTHETIC JOINTS

G.M. Ferguson and C.H. Evans

Department of Orthopaedic Surgery
University of Pittsburgh
986 Scaife Hall
Pittsburgh, PA 15261, USA

In the absence of infection, a correctly inserted prosthetic hip can be expected to function well for several years. However, after about ten years there is a 25 per cent incidence of failure due to aseptic loosening. This probability rises rapidly thereafter, making aseptic loosening the leading cause of the failure of prosthetic joints. Although often considered predominantly a mechanical process, recent evidence points to an important biological component to the aseptic loosening of replacement joints.

Prior to loosening, a radiolucent zone of apparent osteolysis occurs in that area of bone which is immediately adjacent to the implant [1,2]. Furthermore, post-revision inspection of loosened hip stems reveals the presence of a specific membrane growing at the interface between the bone and the polymethylmethacrylate cement. This membrane bears striking histological and biological resemblances to a rheumatoid synovial membrane. Not only does its population of cells include those with the characteristic 'dendritic' or 'stellate' morphology, but it produces collagenase and prostaglandin E_2 (PGE_2)in culture [3]. This 'pseudosynovial' membrane may thus be responsible for the localized bone resorption that occurs in its vicinity, and that may be suspected of potentiating the subsequent mechanical loosening of the implant.

Such a proposal raises two important questions. One concerns the biological origins of the pseudosynovial membrane and the identity of the factors which induce its formation at the surface of the cement. The second concerns the nature of the stimuli which activate the cells of this membrane, provoking their synthesis of collagenase and PGE_2. Continuing, recent research in this laboratory addresses the second question, with particular reference

279

Table 1.   Possible Activators in Synovial
Collagenase Production

Substance	References
Phorbol Myristate Acetate	24
Poly(thylene)glycol	25
Cytochalasin B	4
Lymphokines	11
Monokines (Interleukin 1, Mononuclear cell factor)	12
Collagen	26
Particles of latex, cartilage, various minerals	16,13,17

to the possibility that certain substances used in the construction
of prosthetic joints are responsible for activating the cells of
the pseudosynovial membrane.

THE ACTIVATION OF SYNOVIAL AND PSEUDOSYNOVIAL CELLS

    Cellular and physiological similarities between the synovial
membrane, which naturally lines diarthrodial joints, and the pseudosynovial membrane around failed prostheses, permit the former to be
used in studies of the latter.   Upon appropriate stimulation, synovial cells produce three extracellular neutral metalloproteinases,
collagenase, gelatinase and metalloproteinase III [4,5,6,7].   The
last of these degrades a number of proteinaceous substances, including cartilage proteoglycans, and it is conveniently measured
with casein as a substrate [7,8].   Activated synovial cells also
produce large amounts of PGE_2, whose regulation may be independent
of that of the neutral metalloproteinases [9,10].

    Cultures of normal synovial cells usually produce little colla-
genase or PGE_2, but their synthesis of either or both of these compounds can be provoked by any of a number of activators, some of
which are listed in Table 1.   Synovial cell activation occurs during
many forms of arthritis, where is is of great pathophysiological
significance.   Important physiological activators include lympho-
kines [11], interleukin 1 [12], and wear particles [13] produced by the
mechanical erosion of intra-articular surfaces [14].   Neither of these,
nor any of the other substances listed in Table 1, are likely to be
responsible for in situ activation of the pseudosynovial membrane.
In the search for alternative activators of clinical significance,
the most likely source is the implant itself.   We are directing our

attention to three categories of potential activator.

The first category concerns the surface properties of the implanted materials; the physico-chemical nature of the surfaces may be such that the external membranes of adherent cells are perturbed in a manner producing activation. This possibility draws support from knowing that certain activating agents, such as crystals of monosodium urate, need not be internalized to produce activation[15].

Particulates form the second category of potential activators. Particles of latex [16], cartilage [13], monosodium urate and certain other minerals [17] are known to elicit the production of collagenase by cultured synovial cells. The possibility exists that wear particles of polymethylmethacrylate, metal,plastic, ceramic or other implanted substances can do likewise. Particles of hydroxyapatite have already been shown to do this [18].

Thirdly, we need to consider possible soluble activators such as methylmethacrylate or metals released by dissolution. In this perspective, we have recently identified cobalt as one such soluble mediator [19].

EXPERIMENTS ON THE ACTIVATION OF CULTURED SYNOVIAL CELLS BY COBALT

Among the most common alloys used in the construction of prosthetic joints are those based on cobalt. Vitallium, for instance, has the following approximate composition: Co - 62%; Cr - 31%; Mo - 5%; Ni - 1.5%; Fe - 0.5%. Ex vivo and in vivo studies agree that cobalt is remarkably mobile. Rae [20] has measured soluble cobalt concentrations as high as 100 µg/ml in culture media incubated for 48 hr with particles of cobalt. Pioneering early studies with rabbits, conducted in this laboratory by Ferguson and his colleagues [21], confirmed the transport of cobalt from intramuscular implants to the spleen, liver and kidneys. Such mobility appears to extend to humans, where cobalt is found in the blood, urine and hair of patients with the relevant prosthetic devices [22].

Cobalt is toxic. Insertion of rods of cobalt into developing cartilaginous embryonic rat limbs in vitro prevents growth, kills cells and leads to local resorption of the cartilage [23]. Consistent with this is the observation that phagocytosis of particles of cobalt kills murine macrophages [20], while 10^{-3}-10^{-4} M $CoCl_2$ is toxic towards cultured human lymphocytes (Yamage, personal communications) and embryonic limb cartilage [23].

Despite this, cellular necrosis is not a common feature in patients with prostheses which contain cobalt. However, the possibility remains that sub-toxic doses of cobalt produce more subtle changes in cellular metabolism which are critically relevant to the failure of implants.

To investigate this further, we have examined the effects of cobalt chloride solutions on synovial cell activation. Initial experiments were conducted with an established line of lapine synovial cells, HIG-82. This pseudo-diploid line, derived by spontaneous in vitro transformation, produces large amounts of collagenase, gelatinase, caseinase and PGE_2 upon appropriate stimulation (Georgescu, Mendelow & Evans, unpublished). It provides a convenient means of screening putative activators, with positive findings being subsequently confirmed with primary cultures of lapine and human synovial cells.

Addition of $CoCl_2$ to cultures of synovial cells enhances, by 10-20 fold, the demonstrable collagenase, gelatinase and caseinase activities of their conditioned media. The maximum effect is seen with 10^{-6} M $CoCl_2$ for HIG-82 cells, 10^{-7} M $CoCl_2$ for lapine primary synovial cells, and 10^{-4} M $CoCl_2$ for human cells. Production of the three proteinases progressively declines at concentrations of $CoCl_2$ which exceed these maxima, possibly due to toxicity (Fig. 1). Addition of $CoCl_2$ stimulates by up to 50% the production of PGE_2 by these cells. Thus, the more important effect of cobalt is upon the production of proteinases. Appropriate control experiments confirmed that the apparent effects of the addition of $CoCl_2$ were mediated via cellular activation rather than post-synthetic modification of the enzymes' specific activities or some other indirect mode.

These results are of significance in the context of aseptic loosening only if it can be demonstrated that such activation by cobalt leads to increased bone resorption. There is preliminary evidence from this laboratory (Boniface, Ferguson and Evans, unpublished) that this is so. In the presence of both 10^{-6} M $CoCl_2$ and HIG-82 cells, the in vitro resorption of radiolabelled newborn mouse bones is enhanced 1.5-2 fold (data not shown).

DISCUSSION

Cobalt diffuses from prosthetic joint components to the surrounding tissues. The possibility that it subsequently provokes local osteolysis and consequent biomechanical loosening clearly deserves further study. This is particularly pertinent in view of the move towards cementless fixation, where the bone is in intimate, direct contact with a far greater surface area of the implanted metal and thus is especially vulnerable to its effects.

We urgently need to know precisely the concentrations of cobalt present in the areas immediately adjacent to implants and the molecular nature of the metal. Cobalt will undoubtedly be bound to proteinaceous components, both in bone and our tissue culture experiments, but nothing is known of these ligands. It would also be rewarding to learn more of the biochemistry of the synovial

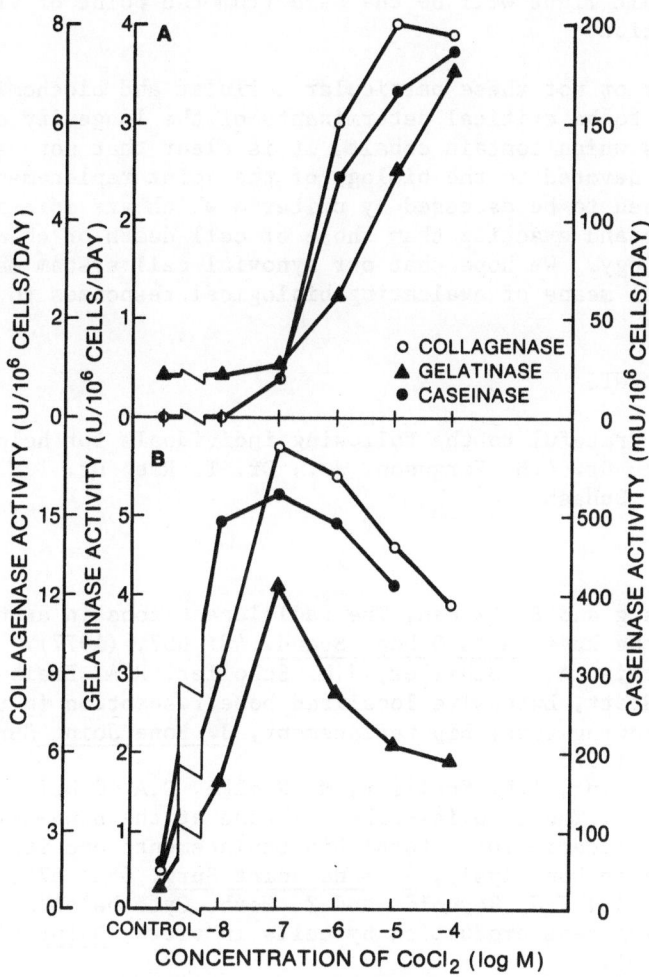

Fig. 1. Effects of $CoCl_2$ upon the production of neutral proteinases by cultures of human (A) and lapine (B) synovial
cells. One unit of proteinase activity degrades 1 µg/min
of the appropriate ^{3}H-labelled substrate at 37°. An activator of latent proteinases, aminophenylmercuric acetate,
was included in each assay.

cells' response to cobalt. For instance, these cells produce an inhibitor of neutral metalloproteinases, known by the acronym TIMP [7]. The possibility exists that cobalt lowers the synthesis of TIMP rather than elevates the synthesis of proteinases, although the end result might well be the same from the point of view of bone resorption.

Whether or not these particular cellular and biochemical activities prove to be critical determinants of the longevity of prosthetic devices which contain cobalt, it is clear that more attention needs to be devoted to the biology of the joint replacement. Tissue responses need to be assessed by criteria which are more subtle, quantitative and exacting than those of cell death or changes in cell morphology. We hope that our synovial cell system may provide ab additional means of evaluating biological responses to orthopaedic implants.

ACKNOWLEDGEMENTS

We are grateful to the following individuals for helpful discussions: Dr. A.B. Ferguson, Jr., Dr. T. Rae, Dr. T.E. Cawston and Dr. H.E. Rubash.

REFERENCES

1. A. Ahlberg and B. Linden, The radiolucent zone in arthroplasty of the knee, Acta Othop. Scand. 48: 687, (1977).
2. W.H. Harris, A.L. Schiller, J.M. Scholler, R.A. Freiburg and R. Scott, Extensive localized bone resorption in the femur following total hip replacement, J. Bone Joint Surg. 58A: 612, (1976).
3. S.R. Goldring, A.L. Schiller, M. Roelke, D.A. O'Neill and W.H. Harris, The synovial-like membrane at the bone-cement interface in loose total hip replacements and its proposed role in bone lysis, J. Bone Joint Surg. 65A: 575,(1983).
4. E.D. Harris, J.J. Reynolds and Z. Werb, Cytochalasin B increases collagenase production by cells in vitro, Nature 257:243 (1975).
5. E.D. Harris and S.M. Krane, An endopeptidase from rheumatoid synovial tissue culture, Biochem. Biophys. Acta. 258: 566 (1972).
6. C.H. Evans and J.D. Ridella, An evaluation of fluorometric proteinase assays which employ fluorescamine, Anal. Biochem. 142: 411 (1984).
7. G.J. Cambray, G. Murphy, D.P. Page-Thomas and J.J. Reynolds, The production in culture of metalloproteinases and an inhibitor by joint tissues from normal rabbits and from rabbits with a model arthritis, I. Synovium, Rheumatol. Internat. 1: 11, (1981).

8. W.A. Galloway, G. Murphy, J.D. Sandy, J. Gavrilovic, T.E. Cawston and J.J. Reynolds, Purification and characterization of a rabbit bone metalloproteinase that degrades proteoglycan and other connective tissue components, Biochem. J. 209: 741, (1983).

9. J.M. Dayer, S.M. Krane, R.G.G. Russell and D.R. Robinson, Production of collagenase and prostaglandins by isolated adherent rheumatoid synovial cells, Proc.Natl.Acad.Sci.USA. 73: 945, (1976).

10. R.M. McMillan, C.E. Brinckerhoff and E.D. Harris, Collagenase release from synovial fibroblasts: relationship to fatty acid release and prostaglandin synthesis, in Inflammation Mechanisms and Treatment, D.A. Willoughby and J.P. Giroud, eds. p.313, University Park Press, Baltimore, (1980).

11. J.M. Dayer, R.G.G. Russell and S.M. Krane, Collagenase production by rheumatoid synovial cells: stimulation by a human lymphocyte factor, Science 195: 181, (1977).

12. J.M. Dayer, S.M. Krane and S.R. Goldring, Cellular and humoral factors modulate connective tissue destruction and repair in arthritic diseases, Sem. Arthritis Rheum. XI suppl. 1: 77, (1981).

13. C.H. Evans, D.C. Mears and J.L. Cosgrove, Release of neutral proteinases from mononuclear phagocytes and synovial cells in response to cartilaginous wear particles in vitro, Biochem. Biophys. Acta. 677: 287, (1981).

14. C.H. Evans, D.C. Mears and J.L. McKnight, A preliminary ferrographic survey of the wear particles in human synovial fluid, Arthritis Rheum. 24: 912, (1981).

15. E.D. Harris, C.A. Vater, C.E. Brinckerhoff, R.M. McMillan and P. Hasselbacher, Patterns of regulation of collagen breakdown in articular cartilage, Sem. Arthritis Rheum. XI suppl. 1: 69, (1981).

16. Z. Werb and J.J. Reynolds, Stimulation by endocytosis of the secretion of collagenase and neutral proteinase by rabbit synovial fibroblasts, J. Exp. Med. 140: 1482, (1974).

17. P. Hasselbacher, R.M. McMillan, C.A. Vater, J. Hahn and E.D. Harris, Stimulation of secretion of collagenase and prostaglandin E_2 by synovial fibroblasts in response to crystals of monosodium urate monohydrate: a model for joint destruction in gout. Trans. Assoc. Amer. Phys. XCIV: 243, (1981).

18. D.J. McCarty, H.S. Cheung, P.B. Halverson and J.C. Garancis, "Milwaukee Shoulder" syndrome: Microspherules containing hydroxyapatite, active collagenase and neutral protease in patients with rotator cuff defects and glenohumeral osteoarthritis, Sem. Arthritis Rheum. XI suppl. 1: 119 (1981).

19. G.M. Ferguson, H.I. Georgescu and C.H. Evans, Cobalt stimulates the synovial production of collagenase and other neutral proteinases, In Press (1985).

20. T. Rae, A study of the effects of particulate metals of ortho-
    paedic interest on murine macrophages in vitro. J. Bone
    Joint Surg. 57B: 444, (1975).
21. A.B. Ferguson, P.G. Laing and E.S. Hodge, the ionization of
    metal implants in tissue, J. Bone Joint Surg. 42A: 77,
    (1960).
22. R.F. Coleman, J.H. Herrington and J.T. Scales, Concentration
    of wear products in hair, blood and urine after total hip
    replacement, Brit. Med. J. 1: 527, (1973).
23. H. Gerber and S.M. Perren, Evaluation of tissue compatibility
    of in vitro cultures of embryonic bone, in: Evaluation of
    Biomaterials., G.D. Winter, J.L. Leray and K. DeGroot eds,
    p.307, John Wiley and Sons Ltd. London (1980).
24. C.E. Brinckerhoff, R.M. McMillan, J.V. Fahey and E.D. Harris,
    Collagenase production by synovial fibroblasts treated with
    phorbol myristic acetate, Arthritis Rheum. 22: 1109, (1979).
25. C.E. Brinckerhoff and E.D. Harris, Collagenase production by
    cultures containing multinucleate giant cells derived from
    synovial fibroblasts, Arthritis Rheum. 21: 745, (1978).
26. C. Biswas and J.M. Dayer, Stimulation of collagenase production
    by collagen in mammalian cell cultures, Cell, 18: 1035,
    (1979).

EVALUATION OF THE HISTOLOGIC RESPONSE TO A POLYACRYLAMIDE-AGAR HYDROGEL IN THE RAT

C.A. Behling and M. Spector

Emory University School of Medicine
Departments of Orthopaedics and Pathology
69 Butler Street
Atlanta, Georgia 30303, USA

Hydrogels, hydrophillic polymers capable of binding large amounts of water [1], have found application in the fabrication of medical devices such as vascular prostheses, contact lenses and wound covering materials. As the use of hydrogels has increased, it has become clear that our understanding of the cellular and tissue responses to this group of hydrophillic polymers is incomplete. Since the functional performance of implants is influenced by the cellular response to the material of fabrication as well as implant configuration, knowledge of cellular reaction elicited by hydrogels will provide for a more rational device design and use.

Hydrogels include acrylimide, methacrylamide, N-vinylpyrrlidone, N-substituted methacrylamides, N-substituted acrylamides, esters of methacrylic acid and others [2]. These materials have water contents up to about 50%. The hydrogels with the highest water contents generally have a relatively low strength that precludes their use for most medical applications. However, a new polyacrylamide-agar composite hydrogel (Geliperm, Geistlich-Pharma, Wolhusen, Switzerland) has relatively high strength whilse comprising 97% water, the highest water content of any hydrogel.

The purpose of this study was to qualitatively and quantitatively characterize the in vivo tissue and cellular resposne to this polyacrylamide-agar (PAA) composite material.

MATERIALS AND METHODS

Strips (3 x 3 x 10 mm) and about 0.5 ml granules (0.5 to 1.5m) of PAA hydrogel, approximately 97% water (by weight), were implanted intramuscularly and subcutaneously into rats for periods of 2-3 weeks

287

and 8 weeks. Surgical grade ultra-high molecular weight polyethylene (PE), (Howmedica, Rutherford, N.J.) strips were used as a
control material. Postmortem specimens were dehydrated in an
ascending series of alcohols and embedded in glycol methacrylate
and ERL embedding medium in preparation for light and electron
microscopy. The mean percent of surface covered by various cell
types was quantitated using a computer interactive image analysis
system.

RESULTS

PAA and PE specimens obtained after 2-3 weeks and 8 weeks were
enveloped by a thin layer of fibrous tissue from 50-200 mm thick
(Figure 1a and 1b). At the observed implant times macrophages
covered about two-thirds of the surface of the PAA with the remaining
surface covered by giant cells and fibrous tissue (Tables 1 and 2).
Sheets of polyethylene control material were covered by macrophages,
fibrous tissue and giant cells in proportions similar to PAA (Tables
1 and 2). Although the amount of surface covered by macrophages
did not change significantly from 2 weeks to 8 weeks slightly more
fibrous tissue appeared adjacent to the sheets implanted for 8 weeks.

In most specimens examined, the material remained intact within
the tissue section. However, when the PAA pulled away from the
tissue section during processing a single layer of cells often
remained attached to the PAA (Figure 2). Such was not the case
with the PE specimens.

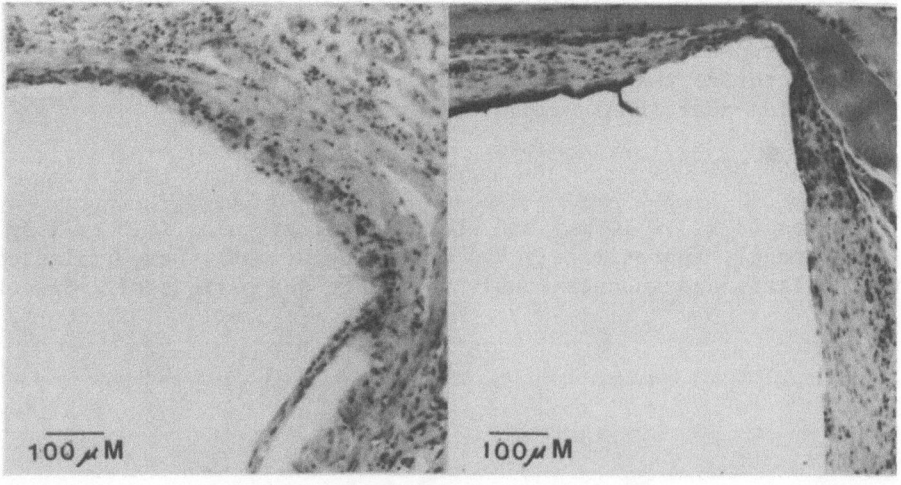

Figure 1a: Fibrous capsule surrounds PAA sheet.
       1b: Fibrous tissue and macrophages appear adjacent to PE
              control material.

Table 1:   QUANTITATIVE EVALUATION OF THE 2 WEEK
             CELLULAR RESPONSE TO PAA AND PE.

	Percent Surface Covered by Cell Type		
	Macrophage	Giant Cell	Fibrous Tissue
PAA	68.37 (26.73)*	27.592 (30.40)	4.04 (5.64)
PE	61.82 (14.43)	35.21 (13.20)	1.79 (2.43)
	NS	NS	NS

PAA - Polyacrylamide - agar hydrogel
PE   - Polyethylene
*   - Standard Deviation
NS   - Not Significant

Table 2:   QUANTITATIVE EVALUATION OF THE 2 MONTH CELLULAR
             RESPONSE TO PAA AND PE

	Percent Surface Covered by Cell Type		
	Macrophage	Giant Cell	Fibrous Tissue
PAA	62.23 (22.71)*	12.18 (11.20)	23.63 (24.62)
PE	49.97 (18.95)	26.06 (18.95)	23.63 (14.28)
	NS	NS	NS

Electron microscope examination of PAA implanted for 8 weeks revealed cells closely apposed to the PAA (Figure 3). Cell processes and collagen fibers were often found extending into PAA material.

Cell debris rarely appeared at the cell/material interface, but was present in deeper cell layers in some sections (Figure 4).

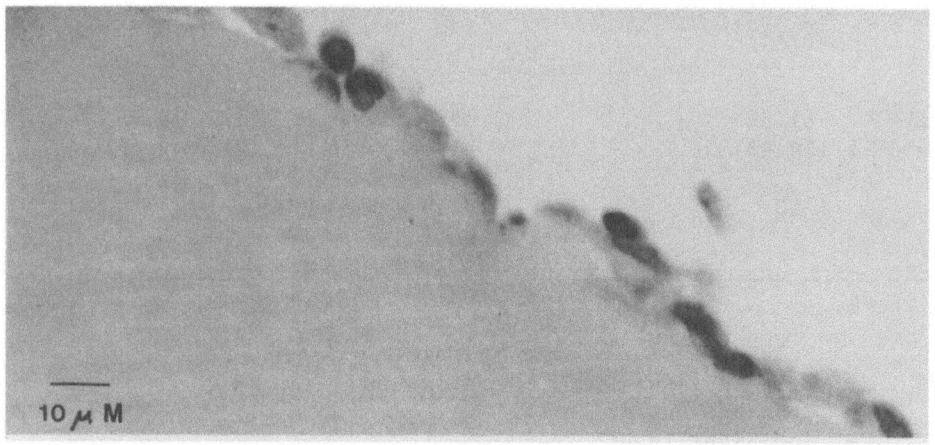

Figure 2:   Cell monolayer adheres to the surface of PAA material which has separated from tissue.

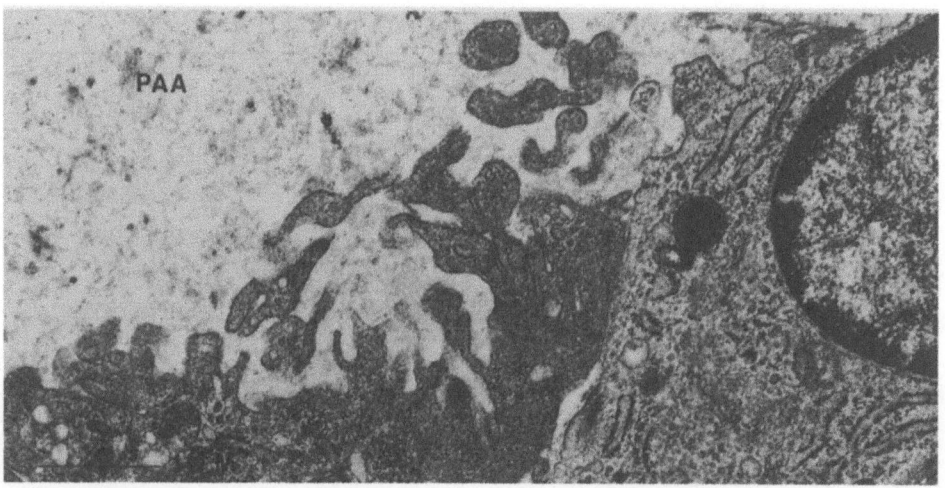

Figure 3:   Macrophage cell processes extend into the PAA material.

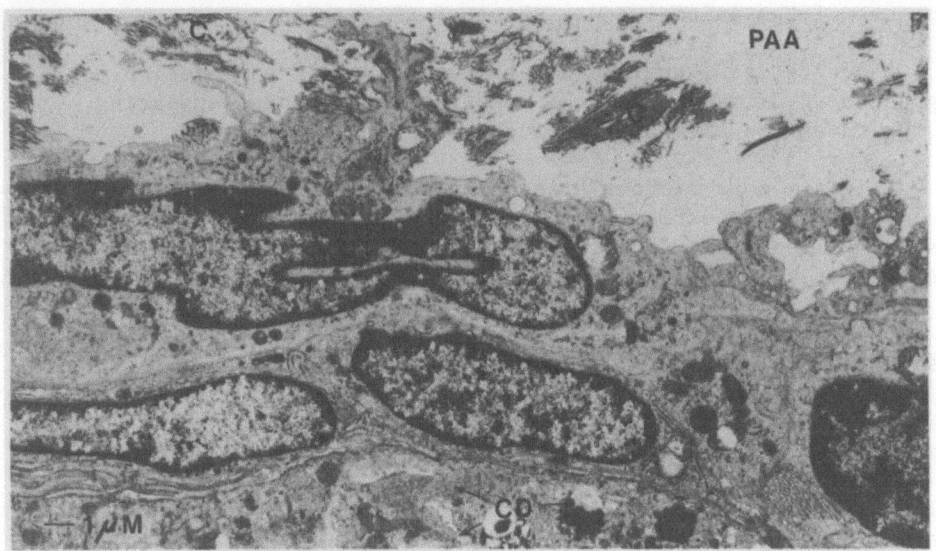

Figure 4:   Collagen bundles (C) appear within the PAA matrix,
             and between cells, and cell debris (CD).

     Cells at the interface possessed features of macrophages,
including numerous cell processes and some subplasmalemmal linear
densities characteristic of mesenchymal cells. Cells with elongaged
nuclei, identified as fibroblasts, also contained dilated rough
endoplasmic reticulum and were surrounded by collagen fibers.
Characteristic fibroblast-like cells were not usually located next
to the implant, although collagen fibers were present at the
surface and within the PAA (Figure 3). Unidentified inclusions
similar to myelin figures are rarely found in the nuclei and cytoplasm of cells at the interface of the hydrogel.

DISCUSSION

     While the co-efficient of variation was high, quantitative
analysis of the cells at the interface of PAA and PE indicate no
significant difference in response to these materials after 2 to
8 weeks. This result was not totally expected due to the different
chemical and physical nature of the polymers. PAA is a complaint
hydrophillic polymer, comprising 97% normal saline, while PE is a
more rigid, hydrophobic polymer.

     The initial reaction to an implant is acute inflammation,
which resolves into a chronic response [3,4]. The cells appearing
at the interface of the PAA and PE characterize the long term
reaction to the materials as a type of chronic inflammatory response,
the foreign body granuloma. Rare polymorphonucleawr leucocytes
(less than 5 per section) in the two week implants are probably

remnants of the initial acute phase of the response.

Quantitative light microscope comparison of each cell type suggested no difference in the number of macrophages appearing at the interface after 2 to 8 weeks.

Slightly more fibrous tissue appeared adjacent to the surface of both the PAA and PE after 8 weeks. This increase in fibrous tissue (collagen and fibroblasts) is probably due to normal progression of the wound healing process.

Quantitative data in this study supports previous qualitative observation of various implanted hydrogels [5,6].

CONCLUSION

The results indicate the tissue compatibility of PAA and illustrate ultrastructural details about the PAA-tissue interface. There is no significant difference between the cellular response to PAA and the surgical grade, ultra high molecular weight PE control material after 8 weeks of implantation in the rat. Macrophages, fibrous tissue and some giant cells were found on the surface of both materials.

REFERENCES

1. B.D. Ratner and A.S. Hoffman, Synthetic hydrogels for biomedical applications, ACS Symposium series No. 31. 1, (1976).
2. L. Sprincl, J. Vacik, J. Kopecek and D. Lim. Biological tolerance of Poly (N-Substituted Methacrylamides). J.Biomed Mater. Res. 5: 197-205, (1971).
3. D.L. Coleman, R.N. King and J.D. Andrade, The Foreign Body Reaction: A chronic inflammatory response, J. Biomed.Mater. Res. 8: 199-211, (1974).
4. B.C. Hirsh and W.C. Johnson, Concepts of granulamatous inflammation, Intl. J. Dermatology 23: 90-100, (1984).
5. S.J. Gourlay, R.M. Rice, A.F. Hegyelli, C.W. Wade, J.G. Dillion, H. Jaffe and R.K. Kulkarni, Biocompability testing of Polymers: In vivo implantation studies. J. Biomed. Mater.Res. 12: 219-232 (1978).
6. R. Langer, H. Brem and D. Tapper, Biocompability of Polymeric delivery systems for macromolecules. J. Biomed. Mater.Res. 15: 267-277, (1981).

CELLULAR INTERACTIONS WITH DENTAL MATERIALS

D.F. Williams

Institute of Medical and Dental Bioengineering
University of Liverpool
Liverpool L69 3BX, U.K.

If one were asked which, of all the classes of biomaterials,
were the most commonly used and had the longest and most extensive
history of such use, few would fail to suggest dental materials as
the answer. On the other hand when asked which class of biomaterials
is associated with the greatest understanding from the biocompatability point of view, few would respond with that same category.
Some of todays dental materials had their origins 100-150 years ago
but we have a very limited perspective on their interactions with
the oral tissues.

When requested therefore to present an overview of the cellular
interactions with dental materials one is left with a dilemma. How
can one talk knowledgeably and authoritatively about a subject which
is so obviously important, when our understanding about it is so
restricted. This difficulty is compounded in the case of this
meeting since we have seen examples of some very elegant work in
cell-cell and cell-surface interactions which have few analogues
in the dental area. Furthermore dentists and dental material
scientists are conspicuous by their virtual absence and one might
conclude that cellular interactions with dental materials are less
important and less significant than is the case with the interactions
between cells and other biomaterials. I do not believe this is the
case and would like to address the reasons for this apparent lack of
understanding or lack of interest, in this overview. It would, of
course, be possible to confine this review to a discussion of the
in vitro cytotoxixity testing of certain types of dental materials
or possibly to a discussion of biocompatability phenomena in dental
implants where a reasonable amount of data exists. In both of these
cases however there are severe limitations to the interpretation of
the data. It would be more appropriate therefore to review some

293

of the features of the biocompatability of dental materials and to
put these into the context of some of the difficulties we face in
this situation.

One of the main difficulties is that there are so many
different kinds of dental materials and different situations in
which they are used. It is an invidious task trying to draw
together information concerning all of these materials and all of
these clinical applications under the one heading of biocompatability
in dental materials. We find, for example, that metals, alloys,
polymers, resins, rubbers, ceramics, composites and natural materials
are all included in the list of commonly used dental materials.
One just has to consider the fundamental differences between
dental amalgam, gold alloys, acrylic resin, silicone rubber,
porcelain and wax to appreciate the difficulties of this task.
Some of these materials may come into contact with pulp, dentine,
enamel, alveolar mucosa during their use. The duration of this
contact may range from a minute or so up to an indefinite time.
There may also be extra-oral contact with, for example, the skin
of either the patient, the dentist or his assistant or even the
gastro-intestinal tract of a patient who should inadvertently
swallow a small component of dental material. In contact with
all of these types of tissue there is the possibility of interactions between these dental materials with a wide variety of cells
including fibroblasts, odontoblasts, epidermal cells and osteocytes.
The scope for material-cell interactions is therefore enormous.

Our problems arise because teeth are inherently susceptible
to destructive disease processes so that they may eventually require
some form of reconstruction. We, of course, do not help matters
very much by our consistent lack of attention to the requirements
of teeth. Certainly our approach to oral hygiene is improving
considerably but we still allow plaque and calculus to accumulate
on the surfaces of teeth and the surrounding tissue, which will
accelerate, and indeed initiate, the destruction of tooth substance and of the peridontal tissues. It is important to recognise
that these conditions exist, and that dental materials that are used
for restorative or reconstructive procedures within the mouth will
have to survive in the same hostile environment.

Several other fundamental problems exist. First, it is
important to recognise that many of the materials we use have to
be in a reactive form when placed within the mouth and indeed
materials have to be adapted to fit each patient. There is no
such thing as a standard dentition or a standard tooth and therefore
in the majority of restorative procedures the material has to be
prepared by the dentist to fit the individual. In some cases the
material is processed in the dental laboratory and subsequently
fitted to the patient; this could be, for example, a crown, a
bridge or a denture. In other cases the material has to be

adapted to some cavity within a tooth. In these cases the material has to be in the form of a viscous fluid for adaptation within the cavity or root canal. Once placed in this position the material has to set, usually to a hard durable solid. This implies that some setting reaction must take place, in itself implying the presence of reactive species. This type of system places a very serious constraint upon the assessment of the biocompatibility of the starting materials which could be in the form of pastes, powders, or liquids? Or is it appropriate to test the properties of the material once it has set? In the former case it is indeed these materials which the tissues will contact initially. However, this will only be a transient contact and since these materials are likely to be highly reactive to tissues the almost certain harmful effect that will be seen in some cases may not be a fair reflection on how that restorative material will influence the tissues in the real clinical situation. On the other hand, in the second case the complete avoidance of those initial reactive substances may again present an unfair bias in the biocompatibility testing, in this case by deliberately ruling out such harmful effect.

We then have to consider the hostile nature of the environment to which these materials may be subjected. Reference has already been made to this situation in the context of the patient's oral hygiene. It is not widely appreciated that the environment of the mouth is very aggressive toward synthetic materials from chemical, biochemical, microbiological and mechanical points of view. We thus find a wealth of literature concerning the corrosion, degradation and erosion of a wide variety of dental materials within the oral environment. We have to take such phenomena into account when considering the effects of the materials themselves on oral tissues since the release of particulate or soluble corrosion products may potentiate, in the medium and long term, the tissue damage that is produced by the materials.

Then we have to consider the specific features of certain of the cells involved. Some of the tissues coming into contact with dental materials are very sensitive to outside interference. The pulp is perhaps the obvious example, being a very sensitive tissue involving many components including mesenchymal cells and fibroblasts. The pulp-dentine border is lined with the odontoblasts whose processes extend into the dentinal tubules. This tissue and these cells in particular are easily damaged and we find that restorative materials themselves, the methods of cavity preparation and the bacteria which will inevitably gain access to the base of the cavity all may have a significant influence on the cells of the pulp, even if the cavity itself does not extend to the pulp-dentine border. This question is discussed in further detail below but at this point it should be recognized that the pulp and indeed other oral tissues are easily damaged and that it is not only the chemical nature of the material which is involved in the production of this damage.

There are several areas of controversy and significance within the subject of the biocompatability of dental materials and it is relevant to consider some of these briefly at this point.

First on this list must come the question of mercury toxicity associated with the use of dental amalgam. Dental amalgams have been used as restorative materials for over 150 years. Throughout this period of use the question of mercury toxicity has always remained controversial with periodic character to the pattern of aggression shown by the anti-amalgam lobby. For much of the time the arguements have centred around the potential danger to the dentist and his assistant on the basis that the air in the dental surgeries contains more mercury vapour than any general environment and that blood and urine analysis of these personnel usually show elevated mercury levels. There is indeed incontrovertible evidence that very poor standards of mercury hygiene in the clinics will lead to excessively high mercury levels which can be dangerous. However, such occurrences are few and far between and normal clinical practice should not lead to any undesirable effects. More recently the pressures exerted upon the dental profession to eliminate mercury based substances from use has increased and has been associated with suggestions that mercury toxicity becomes manifest in patients as well as dental personnel. This has been especially seen in the United States where arguements have been put forward that mercury toxicity arising from dental use is a major contributory factor in a wide variety of diseases including serious conditions such as multiple sclerosis and arthritis, as well as a plethora of less significant conditions such as headaches and general malaise. However, the vast majority of these arguements are based on anecdotal evidence rather than scientific fact and I believe that it is important to appreciate that these conclusions concerning the biocompatibility of one of dentistry's most traditional materials are not substantiated by the available eveidence.

This increasing concern over the safety of dental amalgam has coincided with a considerable effort towards the development of alternatives to amalgam. This is largely coincidental because although many people have argued that the impetus for seeking alternatives to amalgam lies within this question of toxicity, this is not really the case. Of more widespread concern is the fact that dental amalgam, as a restorative material, does not look like tooth substance. Whilst for many years the aesthetic requirements of restorative materials have been widely appreciated in the context of anterior fillings it has not been thought necessary to place any restrictions on the appearance of these materials within the posterior teeth. However, attitudes are changing and there is considerable desire, especially on behalf of the patient, to have restorations to posterior teeth looking like teeth rather than pieces of metal. We have seen during the lat few years the emergence of a new generation of composite restorative materials which initially,

at least, appear to possess many of the required properties for
posterior restorations. It is necessary. of course, that any such
alternatives to amalgam should biocompatibility properties considerably superior to the amalgam. A considerable amount of work has
therefore been performed on the toxicological and biocompatibility
properties of these new materials, the majority of this work being
performed by the manufacturers themselves. It is inappropriate here
to try to discuss the toxicity testing of such materials but a few
comments on the response of oral tissues and especially pulp to
these materials is relevant.

The first point to make here is that the most effective test
system for studying the effect of these materials on the dentine-
pulp complex involves animal experimentation. A cavity may be cut
in the tooth of an experimental animal, this tooth filled with the
test material and then the tooth examined histologically at some
time subsequent to the placement. This immediately poses a problem
however since the very process of cavity preparation will also
inevitably disturb the dentine-pulp complex. In particular, cavity
preparation disrupts the odontoblast processes which are protruding
through the dentinal tubules. This causes a significant response
from these odontoblasts and both cell movement and the infiltration
of inflammatory cells into the pulp may be seen. It is very important therefore, when examining a test material under these circumstances to assess the damage produced by cavity preparation alone
and to make allowances for this in the interpretation of the
histological data.

Secondly, it is becoming abundantly clear that a significant
contribution to the pulpal response to procedures involving restorative materials are associated with bacterial infiltration. It is
well known that when polymer based materials such as these composites
are placed within a cavity there is a degree of polymerisation
shrinkage that takes place during the setting. This, in the absence
of adhesion to the tooth substance, leads to a microscopic gap opening
up at the interface between restoration and the tooth. Although this
gap may be small, indeed very small, it may be sufficiently large to
allow the ingress of bacteria and fluids into this interface. The
presence of such bacteria will naturally influence the cellular
response to the procedure. This, of course, is in addition to the
effects associated with any bacteria introduced to the base of the
cavity at the time of cavity preparation. Again the possibilities
of the observed effects being associated with bacterial contamination
have to be recognised in the interpretation of the data.

Thirdly, we have to consider the response of the dentine as well
as that of the pulp and how this response of dentine influences the
pulp in the long-term. Cavity preparation is a traumatic procedure
and involves the elimination of superficial tissue. We could compare
the situation of a cavity within a tooth to that of an open wound on

epidermal tissues. In the latter case there is well defined mechanism of epidermal healing which will result in the formation of new
tissue. In the case of the cavity in a tooth such healing cannot
take place because enamel is acellular and possesses none of the
properties of epidermal tissues. There is some response of the
dentine to this trauma, however, which results in further mineral-
isation of the dentine, in effect producing secondary or reparative
dentine. If we examine histologically the dentine that remains
between a cavity and the pulp we will usually find the development
of a layer of this reparative dentine. Since the effects of restorative material in that cavity are likely to be associated with the
transport of components of the restorative material through the
dentine to the pulp, the presence of a more sclerotic area will
inhibit that transport and therefore reduce the effects of the
material on the pulpal tissues. There is therefore a slightly
paradoxical situation in which cavity preparation and the initial
presence of an irritant restorative material will cause this response
in the dentine, which over a period of time tends to protect the pulp
from further exposure to irriatation. The time course of events
therefore in studying the histological response of the pulp-dentine
complex to restorative materials may be a little different than that
you would expect from other tissues and the protective effect of
this increasingly sclerotic dentine has to be taken into account.

It can be seen from these three points that the biocompatibility
testing of dental materials is indeed difficult. All of these
factors not only affect the response in the experimental situation
but obviously influence the response in the real clinical world. It
is therefore appropriate to test materials under conditions which
simulate the clinical situation, but this does lead to enormous
problems in the interpretation of the data and in extracting useful
information concerning the direct effects of the material itself.

With these thoughts in mind we should be aware that this new
generation of composite restorative materials appear at this stage
to be minimally irritant to pulpal tissues and so far no evidence
has emerged to suggest that these materials are associated with any
adverse response in this situation.

Moving away from composite materials we find very considerable
academic and industrial interest at this time in the use of alloy
systems in dentistry. For many years gold alloys have been widely,
and in some cases exclusively used, in restorative dentistry for
crown and bridge work. The economics of dental gold alloys have
indicated that alternative alloy systems might be more appropriate.
This has led to the search for base metal alloy systems that would
have the same or closely similar properties to the gold alloys when
used as castings for dental use. There are several limitations to
the use of base metal alloy systems in this context but one of the
most significant concerns certain questions relating to the

biocompatibility of the metals involved. Alloy sytems which are
going to have the desirable combination of mechanical properties and
corrosion resistance coupled with an ease of casting and the ability
to bond to porcelain are all too frequently based on metals such as
cobalt, nickel and chromium. Whilst certain of these alloy systems,
for example the cobalt chromium alloys have a long and successful
history of use both in dentistry and other areas of surgery, there
are undoubtedly problems facing the use of many of these alloys.
The problems seem largely to arise from alloys containing nickel
since metallic nickel produces a significant response from tissues
when implanted and since nickel is a known sensitizer and is frequently associated with hyper-sensitivity responses. There is therefore a flurry of interest at the moment in identifying whether nickel
or indeed any other base metal is potentially harmful in this situation. The evidence is far from clear. It would seem that patients
can indeed be sensitized by nickel and patients fitted with appliances
constructed from alloys containing nickel may display symptoms of an
allergic response. However, there is also evidence that exposure to
nickel in one situation may protect patients from sensitivity responses in other situations and it is by no means certain that dental
patients run a greater risk of hyper-sensitivity responses merely by
being fitted with a nickel containing appliance. This is an area
where a considerable amount of work needs to be performed if we are
to understand these interactions.

It may seem strange that the dental profession chooses to use,
in certain of its restorative materials, metals which are known to
have suspect toxicological properties. The mercury and nickel already
mentioned provide good examples. This phenomenon is not confined
to dental alloys however. In many dental preparations, some of which
have their origins very many years ago, we find strange mixtures of
potentially irritant materials and substances. Asbestos has been a
component of peridontal dressing materials. Cadmium has been a
component of some denture based materials. One of the more important and controversial areas where known irritants are used is that
of cements and sealants used in the root canal. In this situation
it may be important to ensure the removal of all traces of infection.
Whilst the procedures themselves are aimed at thorough cleansing of
the root canal there are many materials available which aim to
enhance this effect. Some of these cements are indeed very effective in sterilising the root canal. In contact with sensitive
tissues however these cements may induce considerable damage. In
root canal filling the protrusion of these substances beyond the
apex into the peri-apical tissues will almost inevitably lead to
an inflammatory response which will probably have very severe
consequences as far as the tooth is concerned. These effects need
not be confined to the tooth and it is not uncommon to find considerable resorption of the alveolar bone and clinically significant
effects on the nerve tissue within this area. Should some of these
materials be used in other endodontic procedures where they may come

into contact with living pulp then it may easily be demonstrated
that they are able to produce pulpal necrosis very readily. It is
true to say that the biological properties of these cements have
only been considered important in relatively recent years. It is
very clear however that we need to rationalize the use of these
irritant materials in this situation and to prevent the occurrence
of these disastrous sequelae that may follow the inappropriate use
of these materials.

        We may turn our attentions briefly to the use of denture based
materials. For over 40 years now polymethylmethacrylate has been
the material of choice for denture bases throughout the world. The
main reason for this lies not with any mechanical, biological,
physical or chemical property but rather with the ease of fabrication in the context of the preparation of individual dentures for
patients. By mixing a powdered prepolymer with liquid monomeric
methylmethacrylate it is possible to prepare a dough whcih can be
polymerised via the affect of heat on the polymerisation initiator
benzoyl peroxide. One consequence of this type of addition
polymerisation system is that a wide spectrum of molecular weight
is produced. The resulting polymer will contain a measurable
amount of unreacted methylmethacrylate and the presence of this small
amount of monomer has given rise to hypotheses concerning the aetiology of denture stomatis. This is the erythema and odema which occurs
on the oral mucosa in association with dentures and the irritation
can be so uncomfortable as to prevent the patient wearing the
denture. It has been commonly thought that this denture stomatitis
is predominantly due to either a direct irritation of the oral
mucosa produced by the residual monomer or alternatively to a hypersensitivity response to this substance. If this were true then
this would be a classical situation of bio-incompatibility of dental
materials. This hypothesis gained some ground when it was realised
that the simple expedient of remaking the denture, during which
process it could be ensured that the amount of residual monomer was
reduced by prolonged curing, resulted in a resolution of the stomatitis. However, it has now become clear that other factors are
involved. indeed the role of the methymethacrylate is probably
minimal and certainly far less significant than that associated with
mechanical trauma to the oral mucosa produced by an ill-fitting
denture, or more importantly to fungal or yeast contamination of
denture surfaces which results in this irritation and inflammation.
Naturally carefully remaking the denture would probably result in a
better fitting and certainly one which was less contaminated. This
brief discussion serves to highlight the multi-factorial character
of phenomena associated with the interactions between dental
materials and oral tissues.

        Mention has already been made of the fact that dental materials
have to operatewithin an oral environment which is hostile, and they
have to experience the same destructive conditions that have

affected the original tissues. One aspect here which is now per-
ceived to be of considerable significance is the role of the adsorbed
proteins, bacteria and acquired pellicle on dental materials in the
mouth. For example, when considering the biocompatibility of a
composite restorative material it is now clear that we have to
consider the response of gingival tissues as well as that of the
pulp. It is unlikely that a fully cured composite material will
chemically irritate the gingival tissues. However, should the
restoration be placed close to the gingival margin, and should the
restoration have a rough surface, then bacterial plaque will
accumulate on that surface and this plaque may and indeed probably
will influence the adjacent gingival tissue. Similarly, plaque
accumulation, or at least, association, with other materials within
the mouth, including dentures and orthodontic appliances, may
influence the oral soft tissues that are in contact with these
appliances. The ability of a dental material to be polished and to
retain a polished finish when subjected to the mechanical and
chemical components of the environment is clearly an important
factor in controlling this plaque accumulation and the subsequent
effect on the tissues.

This brief overview has attempted to highlight some of the
difficulties involved in carrying out biocompatibility studies on
dental materials and in interpreting the data that is derived from
such studies. It has only been during the last decade that the
importance of these biocompatibility phenomena, including reactions
between dental materials and the cellular components of oral tissues,
has been appreciated. There is no doubt that these considerations
will play a very important part in the development of future
generations of dental materials.

SUMMARY AND CONCLUSIONS

CELL-SURFACE INTERACTIONS

SUMMARY AND CONCLUDING REMARKS

D.F. Williams and A. McNamara

Institute of Medical and Dental Bioengineering
University of Liverpool

The first comment to be made is that this conference was a
very brave experiment for which the organisers should be highly
commended. As Dr. Spector said at the conference dinner "The
very nature of the programme itself ensured we could call this
conference a success". Success of course, is a subjective judgement and it may not be until we reflect at length on what was said,
and incorporate such thoughts into our own work that the degree
of success achieved can be determined. However, in a general sense,
the significance of technology transfer in our own inter- and multi-
disciplinary field cannot be over-emphasised and so the very fact
that we are all here representing such diverse backgrounds and
individual preferences provides a sound basis for success.

We have had sessions and papers on three themes. First the
basic cell biology and biochemistry, with discussions on the structure
of cell membranes, electrokinetic properties, the role of proteins
such as fibronectin, and the structure of endothelial surfaces. These
introductory contributions proved to be very impressive and of
particular value to the relative outsiders in these areas.

Then we had discussions on cell-natural surface interactions,
examples of which were the metastatic spread of tumours, platelet
aggregation on the sub-endothelium, the response of epithelium to
injury and the fate of liposomes in the reticulo-endothelial system.

Finally, we had discussions on the biocompatibility of biomaterials which has involved, in some cases although not all, cell-
foreign material interactions. The examples here were the blood
compatibility of materials, phagocytosis of foreign particles, the
influence of wound dressings on epithelial repair, the tissue

305

response and bacterial response to sutures, and specific situations of dental materials interacting with odontoblasts and orthopaedic materials with osteoblasts.

It is not the purpose of this summary to pass comment on the state of the art in these separate areas. What is more relevant is to seek points of commonality within them and to try to relate one with the other. It must be immediately apparent that there are huge gaps, between, for example, experimental cell biology and clinical procedures. One could not help but notice reference made to the difficulty of extrapolating from work on tumours to clinical oncology. The application of basic biological and biochemical knowledge to the use of prosthetic and extra-corporeal devices has proved extremely difficult, as exemplified by the problems of clinical orthopaedics described by Professor Scales. The significance of attempts to bridge these gaps and apply scientific principles was very well demonstrated during the paper given by Dr. Spector.

Having said that the concept of the conference was excellent it would be extremely naive, if we failed to recognise the fact that the very simplistic phenomenological approach widely adopted in biocompatibility studies bears little relation to the wealth of knowledge and the degree of sophistication of experimental techniques that have been demonstrated in basic cell-biology and biochemistry. We cannot hope to progress in biomaterial science until we redress that balance. We have seen some excellent examples of work with this objective but such studies are small in number. If this symposium stimulates just a few more people or groups to recognise this fact, then it may be considered a success.

There are many reasons, of course, why biomaterials science has been retarded in this context, most of these being entirely valid. A paper by Dr. Christopher early in the conference epitomized some of the problems that we currently face. Devices have to be manufactured and the manufacturer has to take the moral and ethical responsibility for securing the safety and efficacy of his device. To many eyes biocompatibility testing has become synonymous with safety testing. The type of test methods discussed by Dr. Christopher, which are widely used in safety testing, may indicate crudely how much of various types of biomaterials my be eaten by rodents before some demonstrable effect or death occurs but reveal nothing about the mechanisms of the interactions between these materials and tissues in the clinical situation. Indeed, should such tests ever reveal facts of any significance, the data would probably remain proprietary, usually only divulged to the regulatory authorities. Millions of pounds per year are spent on testing but the subject of biocompatibility gets no further.

When considering the interactions between biomaterials and tissues we may normally identify four components of these inter-actions:-

1. Initial events at the interface, dominated by protein adsorption.

2. The effects of the environment on the material i.e. corrosion or degradation of biomaterials.

3. The local tissue response to these implanted materials.

4. The systemic manifestations of the above.

What have we learned in respect of these four components from this meeting?

In respect of protein adsorption, the biomaterials community during the last year or so has started to recognise that there are proteins other than albumin and fibrinogen which are capable of interaction with foreign materials and the potential significance of fibronectin has certainly been demonstrated here. We have learned from Dr. Couchman of the ubiquitous distribution of this protein and its role in the promotion of cell adhesion and other cellular pro-cesses including fibroblast and epithelial cell migration. It appears that cyto-skeletal alterations can result from the inter-action between the cell surface and fibronectin and this clearly could be of fundamental importance. It may be, of course, that the apparent significance of fibronectin coincides with the development of techniques for its study, and other proteins may take on equiv-alent status in the future for similar reasons but we certainly cannot ignore the possible influence of fibronectin on biocompat-ibility. The initial events at an interface which involve cells are obviously dependent on their basic structure and biochemistry, and the early papers were extremely useful in providing this back-ground information. Dr. Coleman particularly emphasised the sig-nificant role of the plasma membrane in relation to surface:surface interactions involving cells, tissues and foreign materials. Professor Crawford drew attention to the important changes occuring in the distribution of polar groups within membrane domains which, in addition to more fundamental alterations in the organisation of chemical elements, control basic cellular morphology.

Protein adsorption has been studied quite extensively in relation to the blood compatibility of polymer surfaces and at least we can see in that situation how adsorbed proteins might be involved in the thrombogenic process. It must be said we have to be very careful here not to over simplify the situation and try to ration-alise events in terms of simple physio-chemical parameters; the story is considerably more complex than that.

It is far less easy to identify the role of adsorbed proteins

or indeed other macromolecules on the generalised tissue response
to biomaterials. It is highly probable that we will find great
variability in the nature of protein adsorption on foreign surfaces
but it is not clear by what mechanisms this could control subsequent
events. The comments by Dr. Anderson on cell adhesion to protein
layers on surfaces were of great interest in this respect. One
could imagine with a minimally reactive surface that the very nature
of the adsorbed protein layer might, in the absence of extraneous
factors, initiate a specific and consistent series of cell-surface
interactions. It is interesting here to reflect on the work of the
Gothenburg group referred to by Dr. Spector in which it has been
demonstrated that glycosaminoglycans are present on titanium, or at
least titanium oxide, surfaces and the correlation of deposition of
this layer with osseointegration. The reasons for the occurrence
of such an interaction in this specific case, and indeed the validity
of this observation is not yet clear. We should bear in mind the
discussion concerning the role of surface texture in the control
of the activity of monocytes and how this might relate to initial
events.

We do, however, have a dynamic situation and over a period of
time the subsequent diffusion through the protein layer of corrosion
or degradation moieties and the denaturation and turnover of adsorbed
proteins would obviously influence the sequence of events. There
are many unanswered questions here but hopefully the application
of some of the techniques that have been discussed in this meeting
will help us study these basic interactions.

Little has been said of the influences of cells and their
products on corrosion and degradation and this was perhaps a dis-
appointing feature of the conference. It is well known that the
body provides a very hostile environment which causes alterations
to many implant materials which ultimately prove damaging to the
host. This fact was emphasised by Dr. Anderson who gave examples
of macrophages associated with metal particles found at sites some
distance from corroding implants. Dacron fibres enamating from
heart valves following breakdown may eventually lodge in brain
tissue, while silicone rubber can induce foreign body giant cells
in the liver. The precise nature of the cellular and humoral
environment of a material is very important in determining the rate
of any interfacial reaction that takes place. Anderson has described here his cage implantation system which has been evolved to
examine and quantitate the destructive effects of the body on
implant materials as well as determining the effects of the materials
on the cellular environment. It was interesting to hear the comments
of Dr. Anderson about the pitting of poly(ether urethanes) and its
dependence on the activity of cells, whose presence was induced by
other materials. Bacterial action may also profoundly alter the
properties of biomaterials. Dr. Bucknall described the relationship
between bacteria and suture materials, stating that bacteria may

prevent adsorption of multi-filament sutures, causing them to act
simply as an infected foreign body without aiding wound healing.
Studies have shown that monofilament non-adsorbable sutures are the
most effective for closing infected wounds since they do not harbour
bacteria and retain their strength.

The areas of most interest is the development of the tissue
response and the role of cell-surface interactions in this process.
It is with this subject that the majority of papers were concerned.
Many parameters of the implant material may influence the development of the tissue response. To demonstrate this Professor Taylor
described how changes in surface geometry may induce alterations
to the tissue response both in the context of the soft tissue response and thrombogenicity. In vascular prostheses, for example,
tight dacron weaves induce organised collagen matrices containing
few fibroblasts while velour surfaces cause the formation of thicker
films containing randomly organised collagen. Alterations in surface
chemistry such as the addition of carboxyl radicals to polymeric
vascular implants will also result in decreasing thrombogenesis.
This two way influence also operates extensively at purely biological
interfaces as described by Dr. Pearson. Blood interfaces with
endothelial tissue and a homeostatic mechanism operates here to
control vascular haemostasis. Endothelial cells influence the extent
to which platelet aggregation occurs while the effect of platelet
degranulation causes the endothelial cells to ultimately limit the
degranulation process itself.

It has been interesting listening to the discussion of wound
healing because we may consider the response to an implanted device
as a modification of a wound healing process. Dr. Lang has discussed
how traditional wound management consisted of protection, comfort
and the prevention of infection, often employing the use of anti-
bacterial agents. As discussed by Dr. Brennan however recent work
has shown that antiseptics can be extremely toxic and all have
some degree of deleterious effect on wound healing. Dr. Wokulek
described how changes in the oxygen supply and the degree of surface
drying profoundly altered the healing response. It appears that
moist oxygen-permeable hydrogel-type dressings are very effective,
proving particular advantageous for people who exhibit a slow healing
response. The margin of an open wound is closed by epithelialisation following formation and contraction of granulation tissue.
As described by Dr. Gabbiani, the agents responsible for wound
contraction are myofibroblastic cells now thought to be derived
from fibroblasts in the granulation tissue. Much of wound repair
depends on the deposition of extra-cellular molecules and Dr.
Shuttleworth discussed the growing recognition of the importance
of different types of collagen III and I in scar tissue. As the
scar matures type III is largely replaced by type I. The question
remains, however, as to whether the initially high level of type III
is related to the myofibroblasts at the inflammatory site.

In relation to wound dressings one is always left with the feeling that we are nearly, but not quite, in a new era, with the design of an ideal dressing material and we must look on the present generation of materials with certain expectations.

The degree of organisation of collagenous deposits may also be directly related to the success of an implant material in terms of the tissue response. Dr. Cameron described the implantation of polypropylene mesh during the repair of abdominal wall defects in rats and showed that disorganised deposition of collagen occurred, while a chronic inflammatory response was induced. Repair of the same type of defect with a carbon fibre mesh however, was very successful with organised and oriented collagen layers being produced and with the material acting as a scaffold for its deposition.

The physical characteristics and morphology of the implant surface may be important in controlling the cellular reaction. The importance of elastic modulus imbalance was discussed by Dr. Doyle and related to the acceleration of prosthesis loosening. Dr. Doyle went on to discuss the development of hydroxyapatite reinforced polyethelene composites and their uses in bone replacement. Dr. Spector described the role of porous coatings on orthopaedic materials with the aim of producing an interdigitating bond directly to bone to provide a stable attachment. The importance of extracellular layers deposited on such materials was demonstrated but this has not yet been fully elucidated. The degree of roughness of the particles used in the material preparation seemed to be one factor determining the ratio of the nubmers of macrophages to giant cells within the tissue deposited on the surface.

A lot is known about granuloma formation and it was hoped we would learn something of the relationship of granulomata with the tissue response to biomaterials. Professor Turk told us about the classification of granulomas into immunological and non-immunological types but unfortunately this discussion did not progress very far in relation to the former and to the specific situation of the zirconium and beryllium granuloma that have been so well demonstrated. It would have been interesting to know, for example, why granulomata are seen with zirconium but not with titanium. There is considerable scope for work in this area since the fibrous capsule around an implant could be defined as a special case of a granuloma and since different materials are known to elicit granulomata under different circumstances. These are, of course, complex issues but Dr. Lydon was able to demonstrate how an understanding of the basic processes by which synthetic substrates influence cell behaviour in culture may provide a greater insight into how such interactions occur in vivo. Cell adhesion studies were conducted with a range of polymers and these have shown that increasing hydrophilicity without alteration of chemical groups at the substratum surface will result in

increased cellular adherence. Interestingly cell adhesion was found
to be virtually identical over a wide range of compositions in the
copolymer series, a result inconsistent with any linear dependency
upon chemical group expression. It was suggested that the surface
energy of the material could be extremely important although of
course many other factors are involved.

Unfortunately, severe limitations generally exist in relation
to the interpretation of work performed in vitro. Dr. Rae discussed
the influence of the continuous interaction of cells with the foreign
surfaces that comprise the culture vessels. It is also difficult
to know which changes in the cells to monitor, since many, such as
alterations in cellular morphology, are not completely predictive.
Certainly the physical form of an implant can affect the resulting
response as much as the chemical composition. It was made clear,
however, from the discussion on the response to particulate matter
by Dr. Rae and of the response to porous materials by Dr. Spector
that it is vitally important to adopt the techniques of cell biology
and biochemistry to evaluate these very significant reactions and
the influence of physical and chemical parameters on the nature of
these reactions. The tissue response to implant materials displays
an impressive degree of variability. The response is often simplist-
ically described as fibrous encapsulation, but that term hides a
vast range of micro- and ultra-structural features, the generation
of which may be very specifically related to the characteristics of
the materials in question. Indeed we have been reminded at this
meeting that cell interactions with implants vary in nature from
one part of an implant surface to another. We do not yet know which
material characteristics dominate the sequence of events and it is
sometimes hard to see why small differences in material composition
or surface characteristics should seem to transform completely the
nature of the response. Listening to the discussions on tumours and
their metastatic spread it was interesting to note how many different
types of cell behaviour are possible in relation to micro-environ-
mental conditions. One can visualise, for example, how certain
characteristics of our foreign surfaces could selectively promote the
growth of particular sub-sets of cells.

Studies of the various processes involved in the systemic spread
of tumours provided very good examples of cell natural surface interactions, in which tumour cells interact with platelets on the endothelial or sub-endothelial surfaces. According to Dr. Jamieson,
tumour cells can induce the aggregation in vitro and the same mechanism may be involved in encapsulating daughter tumour cells in the
circulation in vivo. Dr. Post added that the systemic spread of
malignant tumour cells occurs when host defences fail to recognise
and destroy them. It was further suggested by Dr. Langer that early
invasion into adjacent tissues may be promoted by local proteolysis
by the tumour itself but the evidence remains inconclusive at
present. Dr. Post described how experimental approaches involved

in the identification of cellular changes which correlate with the
ability to metastasise, have proved extremely complex. Nevertheless, it appears that alterations in carbohydrate groups on cell
surface proteins can affect the development of neoplasms. An
increase in the accumulation of glycosaminoglycans in the cell
surface of rhabdomyosarcomas, for example, was seen to be inversely
correlated with the metastatic potential of the cells.

Little was said of the systemic sequelae of cell-foreign
surface interaction, probably because far less is known about them
than of the interactions themselves. The discussion by Dr. Westerby
of the results of cardio-pulmonary by-pass and the introduction
here of the term 'total body inflammatory reaction' was perhaps the
most revealing. This descriptive terminology has arisen because
during cardiopulmonary by-pass a systemic inflammatory response
appears to be initiated, following local exposure to foreign
materials. Cells ultimately damage the pulmonary capillaries due
to the release of cathepsins and oxygen free radicals by the neutrophils in that area following reperfusion of the lungs. Severe falls
in plasma fibronectin levels and platelet numbers also occur after
aortic cross-clamping, according to Dr. Powell, which may prove
detrimental to the cellular colonisation of vascular-grafts and
lead to the promotion or generation of micro-emboli.

Some of the presentations have provided us with an insight into
the clinical problems associated with biocompatibility. The discussion on dental materials has, for example, shown how difficult
it is to interpret experimental data derived from biocompatibility
testing of these materials and the review has described some of the
problems involved. Caution must always be exercised in attempting
to obtain the maximum degree of mutual interaction between prostheses
and tissues. The stabilisation of orthopaedic implants by direct
bond ingrowth into porous surfaces is a point in question for
although this appears to be a very suitable answer to problems of
prosthesis attachment, Professor Scales advocates caution on the
grounds that radical surgery would be required should revision ever
become necessary. Obviously this is a controversial area where
opinion will be polarised into two distinct camps. A description
of the development of a small bore microporous polyurethane
vascular prosthesis by Mr. Annis was very interesting since the
nature of the fibres from which the prostheses are constructed
appears to have a significant influence on the development of the
tissue response. The question of the tissue response to vascular
prostheses has occurred several times during this meeting and Dr.
Anderson has demonstrated that the degree of graft maturation which
occurs is dependent on the extent of thrombus formation. This
implies that the configuration of grafts may be extensively altered
by pseudo-intimina hyperplasia.

This summary of the papers and the general discussion shows

that an enormous amount of ground was covered in this conference.
Clearly we have a long way to go before we can hope to explain
some of the clinical and experimental observations concerning the
biological fate of the implanted prostheses in terms of fundamental
cell surface interaction. One must hope, however, that symposia
such as this can only serve to enhance this understanding and to
increase our desire for more knowledge in this area.

BARBENEL, J. C., Bioengineering Unit, University of Strathclyde, Glasgow, Scotland.

BASTIDA, Eva, Hospital Clinico, Barcelona, Spain.

BECKER, M., IRSC du CNRS, ER 278, F-94802 Villejuif, France.

BEHLING, C. A., Emory University School of Medicine, Departments of Orthopaedics and Pathology, 69 Butler Street, Atlanta, Georgia 30303, U.S.A.

BRENNAN, S. S., University Departments of Surgery and Pathology, University of Bristol, Bristol, U.K.

BURNAND, K. G., Departments of Surgery and Histopathology, St. Thomas's Hospital Medical School, London, SE1 7EH.

CAMERON, A., Department of Applied Physiology and Surgical Sciences, Royal College of Surgeons of England, Lincoln's Inn Fields, London, WC2A 3PN.

CAPPERAULD, I., Research Unit, Ethicon Limited, Edinburgh, Scotland.

COUCHMAN, J. R., Biosciences Division, Unilever Research, Colworth Laboratory, Sharnbrook, Bedford. MK44 1LQ, U.K.

COURTNEY, J. M., Bioengineering Unit, University of Strathclyde, Glasgow, Scotland.

CRAWFORD, N., Department of Biochemistry, Royal College of Surgeons of England, Lincoln's Inn Fields, London, WC2A 3PN.

CROCKER, P. R., Departments of Histopathology and Urology, St. Bartholomew's Hospital, West Smithfield, London, E.C.1.

DAVID, P., Centre de Recherches Chirurgicales, CNRS UA 591, Faculte de Medecine, Paris XII, Creteil Cedex 94000, France.

DAVIES, D. R., Departments of Surgery and Histopathology,
   St. Thomas's Hospital Medical School, London, SE1 7EH.

DAVIES, J. E., Department of Anatomy, University of Birmingham,
   Birmingham, B15 2TJ, U.K.

EVANS, B., Departments of Surgery and Histopathology,
   St. Thomas's Hospital Medical School, London, SE1 7EH.

EVANS, C. H., Department of Orthopaedic Surgery, University of
   Pittsburgh, 986 Scaife Hall, Pittsburgh, PA. 15261, U.S.A.

FERGUSON, G. M., Department of Orthopaedic Surgery, University of
   Pittsburgh, 986 Scaife Hall, Pittsburgh, PA. 15261, U.S.A.

FORBES, C. D., Department of Medicine, Glasgow Royal Infirmary,
   Glasgow, Scotland.

FOSTER, M. E., University Departments of Surgery and Pathology,
   University of Bristol, Bristol, U.K.

GABBIANI, G., Department of Pathology, University of Geneva,
   1211 Geneva 4, Switzerland.

HASTINGS, G. W., Bio-Medical Engineering Unit, (N. Staffs
   Polytechnic and Staffordshire Health Authority), Medical
   Institute, Hartshill, Stoke-on-Trent, U.K.

HILL, S. F., Departments of Histopathology and Urology,
   St. Bartholomew's Hospital, West Smithfield, London, E.C.1.

HURST, R. P., Department of Anatomy, University of Birmingham,
   Birmingham. B15 2TJ, U.K.

IRWIN, J., Department of Surgery, Charing Cross Hospital,
   Fulham Palace Road, London, W6 8RF.

JAMES, A., Department of Surgery, Charing Cross and Westminster
   Medical School, Fulham Palace Road, London, W6 8RF.

JAMIESON, G. A., American Red Cross Laboratories, Bethesda,
   Maryland 20814, U.S.A.

LANE, I., Department of Surgery, Charing Cross Hospital,
   Fulham Palace Road, London, W6 8RF.

LAWFORD, P., Department of Medical Physics and Clinical
   Engineering, Royal Hallamshire Hospital, Sheffield, Yorks.
   S10 2JF.

LAYER, G. T., Departments of Surgery and Histopathology,
St. Thomas's Hospital Medical School, London, SE1 7EH.

LEAPER, D. J., University Departments of Surgery and Pathology,
University of Bristol, Bristol. U.K.

LLOYD-DAVIES, E., Department of Surgery, Charing Cross and
Westminster Medical School, Fulham Palace Road, London, W6 8RF.

LOISANCE, D., Centre de Recherches Chirurgicales, CNRS UA 591,
Faculte de Medecine, Paris XII, Creteil Cedex 94000, France.

LOWE, G. D. O., Department of Medicine, Glasgow Royal Infirmary,
Department of Medicine, Glasgow Royal Infirmary, Glasgow,
Scotland.

McCOLLUM, C. N., Department of Surgery, Charing Cross and
Westminster Medical School, London, W6 8RF.

McNAMARA, A., Institute of Medical and Dental Bioengineering,
University of Liverpool, Liverpool, U.K.

MILLER, Carol, Department of Surgery, Charing Cross Hospital,
London, W6 8RF.

MOCZAR, E., Laboratoire de Biochimie du Tissu Conjonctif,
University of Paris XII, 8 rue du General Sarrail, Creteil
Cedex 94010, France.

MOCZAR, Madeleine, Laboratoire de Biochimie du Tissu Conjonctif,
University of Paris XII, 8 rue du General Sarrail, Creteil
Cedex 94010, France.

MYNETT, K. J., Bioengineering Unit, University of Strathclyde,
Glasgow, Scotland.

ORDINAS, A., Hospital Clinico, Barcelona, Spain.

PATTISON, M., Departments of Surgery and Histopathology,
St. Thomas's Hospital Medical School, London, SE1 7EH.

PEARSON, J. D., Section of Vascular Biology, M.R.C. Clinical
Research Centre, Harrow, Middlesex. HA1 3UJ, U.K.

POSKITT, K. R., Department of Surgery, Charing Cross and
Westminster Medical School, London, W6 8RF.

POSTE, G., Smith Kline and French Laboratories, Philadelphia,
Pennsylvania 19101, U.S.A.

POUPON, M. F., IRSC du CNRS, ER 278, F-94802 Villejuif, France.

POWELL, J. T., Department of Biochemistry, Charing Cross and
    Westminster Medical School, London, W6 8RF.

RAE, T., Orthopaedic Research Unit, University of Cambridge
    Clinical School, Addenbrookes Hospital, Cambridge. CB2 2QQ.

RAMSAY, J. W. A., Departments of Histopathology and Urology,
    St. Bartholomew's Hospital, West Smithfield, London, E.C.1.

SHUTTLEWORTH, A., Department of Biochemistry, Manchester
    University Medical School, Stopford Building, Oxford Road,
    Manchester. M13 9PT.

SILVER, I. A., University Departments of Surgery and Pathology,
    University of Bristol, Bristol, U.K.

SINCLAIR, Marion, Department of Surgery, Charing Cross Hospital,
    Fulham Palace Road, London, W6 8RF.

SPECTOR, M., Emory University School of Medicine, Departments of
    Orthopaedics and Pathology, 69 Butler Street, Atlanta,
    Georgia 30303, U.S.A.

SPOONER, N. T., Department of Anatomy, University of Birmingham,
    Birmingham, B15 2TJ, U.K.

TAYLOR, D. E. M., Department of Applied Physiology and Surgical
    Sciences, Royal College of Surgeons of England, Lincoln's Inn
    Fields, London, WC2A 3PN.

TURK, J. L., Department of Pathology, Institute of Basic Medical
    Sciences, Royal College of Surgeons of England, Lincoln's Inn
    Fields, London, WC2A 3PN.

WESTABY, S., Department of Surgery, Cardiothoracic Division,
    Hammersmith Hospital and Royal Postgraduate Medical School,
    London, W.12.

WHITFIELD, H. N., Departments of Histopathology and Urology,
    St. Bartholomew's Hospital, West Smithfield, London, E.C.1.

WILLIAMS, D. F., Institute of Medical and Dental Bioengineering,
    University of Liverpool, Liverpool. L69 3BX, U.K.

WOODS, Anne, Biosciences Division, Unilever Research, Colworth
    Laboratory, Sharnbrook, Bedford. MK44 1LQ, U.K.